Scientific Instruments on Display

History of Science and Medicine Library

VOLUME 46

Scientific Instruments and Collections

STUDIES PUBLISHED UNDER THE AUSPICES OF THE SCIENTIFIC INSTRUMENT COMMISSION

VOLUME 4

The titles published in this series are listed at *brill.com/hsml*

Scientific Instruments on Display

Edited by

Silke Ackermann
Richard L. Kremer
Mara Miniati

TUTA SUB AEGIDE PALLAS · 1683 ·

BRILL

LEIDEN | BOSTON

Cover illustration: The Paris Observatory: the hall of the museum and the meridian line, from *Le Journal Illustré* 16, 9 (1879), p. 68, detail.

Library of Congress Cataloging-in-Publication Data

Scientific instruments on display / edited by Silke Ackermann, Richard L. Kremer, Mara Miniati.
pages cm. — (History of science and medicine library, ISSN 1872-0684 ; volume 46)
(Scientific instruments and collections ; volume 4)
Includes bibliographical references and index.
ISBN 978-90-04-26439-7 (hardback : acid-free paper) — ISBN 978-90-04-26440-3 (e-book) 1. Scientific apparatus and instruments. 2. Exhibitions—Social aspects. I. Ackermann, Silke, editor. II. Kremer, Richard L. (Richard Lynn), editor. III. Miniati, Mara, editor.

Q185.S39 2014
507.4—dc23

2014024624

This publication has been typeset in the multilingual 'Brill' typeface. With over 5,100 characters covering Latin, IPA, Greek, and Cyrillic, this typeface is especially suitable for use in the humanities.
For more information, please see www.brill.com/brill-typeface.

ISSN 1872-0684
ISBN 978-90-04-26439-7 (hardback)
ISBN 978-90-04-26440-3 (e-book)

This book is printed on acid-free paper.

Contents

List of Illustrations

Colour Plates

Figures

Chapter 8, Richard A. Paselk, *The Display of Twentieth-Century Instruments at Humboldt State University*

Chapter 9, Peggy Aldrich Kidwell and Amy Ackerberg-Hastings, *Slide Rules on Display in the United States, 1840–2010*

Chapter 11, Ileana Chinnici, Donatella Randazzo and Fausto Casi, *Instruments on Movie Sets: A Case Study*

Chapter 12, Inga Elmqvist Söderlund, *Display of Instruments on Seventeenth Century Astronomical Frontispieces*

Notes on Contributors

Amy Ackerberg-Hastings
holds a PhD in the history of technology and science from Iowa State University. From 2011 to 2013, she digitized catalogue descriptions for over 650 mathematical instruments and documentation, including slide rules, and organized the records into online exhibits for the Smithsonian's National Museum of American History. She has also written extensively on the use of protractors and other tools in the history of mathematics education, and she teaches historical writing for University of Maryland University College.

Silke Ackermann
joined the British Museum, London, in 1995 as the first specialized curator of European and Islamic history of science and scientific instruments. In early 2012 she left the British Museum to take up a professorship at the University of Applied Sciences in Schwerin (Germany) where she became course director of "Cultural studies in a modern world". In March 2014 she was appointed Director of the Museum of History of Science in Oxford. Since July 2013 she has been President of the Scientific Instrument Commission. She is particularly interested in the transfer of knowledge between the Islamic World and Europe and strongly supports cross-cultural and interdisciplinary approaches in museum work.

Marco Beretta
is Professor of the History of Science at the University of Bologna. He has served as editor-in-chief of *Nuncius: Journal of the Material and Visual History of Science* since 2004. His latest book is *The Alchemy of Glass: Counterfeit, Imitation, and Transmutation in Ancient Glassmaking* (Sagamore Beach, 2009). In March 2013, he received the Paul Bunge Prize, an international award for historians of scientific instruments.

Laurence Bobis
is, since 2001, director of the library and collections at the Paris Observatory. She co-edited with James Lequeux *L'Observatoire de Paris: 350 ans de science* (Paris, 2012) and curated several exhibitions including *Léon Foucault, le miroir et le pendule* (2002), *François Arago et l'Observatoire de Paris* (2003), *"c" à Paris* (2005), *Le Verrier: à la découverte de Neptune* (2011), *Hommage à Jean-Dominique Cassini: l'Astronome du roi et le satellite* (2012).

Alison Boyle

is Deputy Keeper of Science and Medicine at the Science Museum in London. She has responsibility across the physical sciences collections with a particular focus on twentieth-century physics and astronomy. Galleries and exhibitions worked on include *Exploring Space* (2006), *Cosmos & Culture* (2009), *Stephen Hawking at 70* (2012) and *Collider: Step inside the World's Greatest Experiment* (2013). Alongside curatorial work she is pursuing at University College London a PhD in Science and Technology Studies, on collecting and displaying the material culture of modern physics.

Fausto Casi

is Professor of Electronics and Director of the del Centro di Formazione Professionale of Arezzo, Italy. He had long cultivated studies on the history of science and technology and is an active collector of items dealing with the evolution of communication media. He founded the Associazione Italiana per la Radio d'Epoca (AIRE) and has authored many articles, exhibition catalogues and books on historical scientific and technological instruments. In 2005 he was appointed founding director of the Museo dei Mezzi di Comunicazione of Arezzo.

Ileana Chinnici

is currently a research astronomer at the Palermo Astronomical Observatory, after having been Curator of its museum from 1996 to 2004. From 2005 to 2010 she served as Coordinator of the Servizio Musei of the National Institute for Astrophysics (INAF) and from 2006 to 2012 as Chair of the Working Group "Archives" of Commission 41 (History of Astronomy) of the International Astronomical Union (IAU). Her main research topic is nineteenth-century astronomy and astrophysics.

Suzanne Débarbat

spent her career entirely at the Paris Observatory, beginning in 1953. From 1955 she was an *assistant, aide-astronome, astronome adjoint* and *astronome titulaire*, and since 1997 *astronome titulaire honoraire*. She is *Docteur d'Etat* and from 1985 to 1992 directed a CNRS research group "Systèmes de référence spatio-temporels", and from 1987 to 1992 directed the Département d'Astronomie Fondamentale of the Paris Observatory (now named *Systèmes de Référence Temps-Espace*—SYRTE, to which she has been attached since 1997). She is a member of the International Astronomical Union (IAU), Commission 41 (History of Astronomy), of the Bureau des Longitudes, and the Académie Internationale d'Histoire des Sciences. From 1975 she has studied the history of the Paris Observatory, its astronomers and their instruments, works and discoveries.

Richard Dunn
is Senior Curator and Head of Science and Technology at Royal Museums Greenwich. He is currently Co-Investigator on a five-year research project on the history of the British Board of Longitude, funded by the Arts and Humanities Research Council, and has been working on a major exhibition on the longitude story. His previous publications include *The Telescope: A Short History* (2009) and *Re-Inventing the Ship* (co-edited with Don Leggett, 2012).

Inga Elmqvist-Söderlund
is Research Fellow at The Royal Swedish Academy of Letters, History and Antiquities and affiliated with the Department of Art History at Stockholm University. From 2011 to 2013 she worked on a post-doctoral project, *Scientific Instruments in 18th-Century Europe: Display, Visual Effect and Aesthetical Experience*, based at the Museum for History of Science at Oxford. In 2013 she was a Sackler Short-Term Fellow at Royal Museums Greenwich with the project *18th-Century Astronomical Demonstration Instruments and User Experience*. From 1996 to 2013 she was curator and director of the Observatory Museum in Stockholm. She has written on and curated several exhibits on early modern art and science.

Ingrid Jendrzejewski
is an independent scholar interested in the intersections between science and theatre. After receiving degrees in English Literature and Creative Writing from the University of Evansville, Indiana, she took a BSc and MSc in Natural Science (Physics) from the University of Cambridge. She has staged many plays including Ben Jonson's *The Alchemist*, Sir Patrick Moore's *Galileo*, Michael Frayn's *Copenhagen*, Carl Djerassi and Roald Hoffmann's *Oxygen* and Carl Djerassi's *Calculus*.

Peggy Aldrich Kidwell
historian of science, is Curator of Mathematics at the Smithsonian's National Museum of American History, Washington, D.C. Her publications concern historical aspects of astrophysics, mathematical instruments, mathematics education, and computing. Topics of her exhibits range from arithmetic teaching to computer bugs. Her present research concerns the history of

mathematical recreations. She is now working with others to display Smithsonian mathematical instruments in an online encyclopedia.

Richard L. Kremer
teaches history of science at Dartmouth College and curates that university's collection of historic scientific instruments. Each year he teaches a seminar on the material culture of science in which his students create an exhibit from that collection. With David Pantalony and Francis Manasek he authored *Study, Measure, Experiment: Stories of Scientific Instruments at Dartmouth College* (Norwich, 2005). He has also published numerous articles on nineteenth-century physics pedagogy, high-speed cinematography, and early modern astronomy and its material culture.

Mara Miniati
was Vice-Director of the Istituto e Museo di Storia della Scienza in Florence and presently is Curator Emeritus of the Museo Galileo. She served as Editor-in-Chief of *Nuncius* and has authored more than 150 publications. In collaboration with the Istituto Centrale per il Catalogo e la Documentazione (ICCD) of the Italian Ministry for Cultural Heritage, she developed the STS cataloguing format for scientific instruments of historical interest. In March 1993, she received the Paul Bunge Prize. Currently she is President of the Museums Committee of the Regione Toscana.

Richard A. Paselk
Professor of Chemistry at Humboldt State University (HSU), earned a PhD in Biochemistry at the University of Southern California. He is founder and curator of the HSU Scientific Instrument Museum, and curator and webmaster for the HSU Natural History Museum. From 1988 to 1995 he was a principal investigator in chemical education in a large project supported by the National Science Foundation, state and corporate sources. He has more than fifty scholarly and educational publications or presentations and has curated more than thirty small museum exhibits including web exhibits. Current interests include informal education and museums.

Donatella Randazzo
Biologist and Librarian, is currently in charge of the historical library and archives at Palermo Astronomical Observatory. She has taken part in the compilation of the bibliographic dictionary of Italian astronomers of the past 150 years, and is presently engaged in the publication of the historical archives inventory.

Steven C. Turner
is curator of physical sciences at the Smithsonian's National Museum of American History, Washington, D.C. He is a former editor of the journal *Rittenhouse*. His research interests include the history of American science education, the history of physics and the history of astronomy. He has collected extensively in all these areas. He is currently exploring links between early American photography and the growth of American astronomy in the nineteenth century.

Introduction

'Scientific instruments on display' can mean many things for many people—display for sale in shop windows or in trade catalogues; display to compete for attention and prizes at Great Exhibitions and trade fairs; display in private collections to save from destruction, to understand or to show off; display in museums and galleries to explain, to teach and to enlighten; display in public spaces or in manuscripts and books to signal the authority or credibility of science; display on stage and in film to provide a context and to (literally) set the scene. Scientific instruments, that is, have long been shown in spaces quite different from their 'working homes' of the laboratory, the classroom, or the observatory.

There are many reasons to display and many spaces where the display can happen. Displays change according to times, contexts and cultures. Different stakeholders—be it producers, owners, curators or members of the public, to name but a few—have different ideas and ideals. They can prefer a didactical approach or stress the aesthetic value of the instruments; they can focus on multimedia or see historic instruments as the core of an exhibition; they can favour recent scientific achievements over those of the past; they can choose to juxtapose the responses different cultures found to the same questions. Changing displays can thus reflect changing attitudes towards instruments, science and technology, and medial and museological spaces in different societies.

Not much has been published on many of these topics. John R. Milburn and Paolo Brenni have examined the display of instruments in printed trade catalogues, David J. Bryden and Michael A. Crawforth the advertisements on ephemeral imprints like trade cards or broadsides. Jim Bennett and Brenni have reviewed the presentation of scientific instruments at several of the 'universal expositions' held in the nineteenth century. Recently published histories of the Whipple Museum of the History of Science, the Science Museum in London, the Deutsches Museum in Munich, or the Smithsonian Institution in Washington have mentioned the changing presentation of scientific instruments in their galleries. And curators and other scholars have written programmatic suggestions for how museums, more generally, might think about displaying instruments. But no general frameworks have been proposed for thinking about the cultural, technical or scientific significance of how scientific instruments have been displayed in venues other than those for which they were originally made.[1] We hope that the essays in this volume will provide some foundations upon which such frameworks may be erected.

1 Cf. John R. Millburn, *Retailer of the Sciences: Benjamin Martin's Scientific Instrument Catalogues, 1756–1782*, London, 1986; Paolo Brenni, "19th Century Scientific Instrument

The decision to choose 'Displaying Scientific Instruments' as the topic for the 29th Symposium of the Scientific Instrument Commission (SIC) of the International Union for History and Philosophy of Science (IUHPS), held in Florence in 2010, was a reflection of intense discussions within the Commission, prompted by the newly opened Museo Galileo.[2] This museum had just reorganized its displays, both thematically and technically. Arrangement based on context (e.g., research, pedagogy, spectacle) replaced the classical subdivision of the natural philosophical disciplines (e.g., mechanics, heat, optics) and digital displays and applications supplemented the usual museum labels. Since such large-scale reinstallations occur infrequently in major museums, it seemed timely to consider the topic of display for the SIC Symposium meeting in Florence.

There were no restrictions with respect to time, place, culture or context. A wide range of papers was presented on wildly diverse topics. Alas, only a small selection could be published at the end and the editors are fully aware of the lacunae. Nevertheless, we hope that the papers published here will enable the reader to get an overview of the breadth and depth of the discussions in Florence.

In his intentionally double-length opening contribution, Marco Beretta sets the scene both of the conference and of the volume. He focusses on the instrumental role of Andrea Corsini, whose determination and tenacity from the opening years of the twentieth century to conserve the scientific-historical

---

Advertising", *Nuncius* 17 (2002), pp. 497–513; David J. Bryden, "Early Printed Ephemera of London instrument Makers: Trade Catalogues", *Bulletin of the Scientific Instrument Society* 64 (2000), pp. 13–16; Bryden, "Evidence from Advertising for Mathematical Instrument Making in London, 1556–1714", *Annals of Science* 49 (1992), pp. 263–285; Michael A. Crawforth, "Evidence from Trade Cards for the Scientific Instrument Industry", *Annals of Science* 42 (1985), pp. 453–554; Jim Bennett, *Science at the Great Exhibition*, Cambridge, 1983; Paolo Brenni, "La science française au Crystal Palace", in Patrice Bret (ed.), *Les techniques et la technologie entre France et Grande-Bretagne, XVIIe–XIXe siècles*, Paris, 2010, pp. 255–265; Liba Taub, Frances Willmoth (eds.), *The Whipple Museum of the History of Science: Instruments and Interpretations to Celebrate the 60th Anniversary of R.W. Whipple's Gift to the University of Cambridge*, Cambridge, 2006; Peter J.T. Morris, *Science for the Nation: Perspectives on the History of the Science Museum*, London, 2010; Helmuth Trischler, Wilhelm Füssl (eds.), *Geschichte des Deutschen Museums: Akteure, Artefakte, Ausstellungen*, Munich, 2003; Pamela M. Henson, "'Objects of Curious Research': The History of Science and Technology at the Smithsonian", *Isis* 90 (1999), pp. 249–269; Christoph Lüthy, "Museum Spaces and Spaces of Science: Reflections on the Explanatory Possibilities of History of Science Collections", *Nuncius* 20 (2005), pp. 415–429.

2 Filippo Camerota (ed.), *Displaying Scientific Instruments: From the Medici Wardrobe to the Museo Galileo*, Milan, 2012.

heritage of his city and his country finally led to the inauguration in 1930 of the Museo di Storia della Scienza (since 2010: Museo Galileo). Beretta paints a vivid picture of the discussions amongst, and the alliances between, the various stakeholders. It was Corsini himself, as the Museo's first director, who was to shape the displays and the rationale behind new acquisitions for over twenty years.

One might be tempted to assume that three national museums in the same capital city might apply similar criteria regarding the display of scientific instruments. That this is not necessarily the case, rather that the changing attitudes towards these displays can be very closely linked with the changing identities of these major institutions, is exemplified by the second group of papers in this volume. The title 'Science Museum' seems to have first been used in 1877 for the non-art collections of the South Kensington Museum in west London, but an independent entity only came into being in 1909. Alison Boyle discusses, with reference to Florence, the specific challenges that were (and are) posed by having to display collections that lack the visual appeal of art objects. Richard Dunn tells a rather different story, discussing the other part of the Kensington Collections or the Victoria & Albert Museum as it was officially called since 1899, by examining instruments considered as applied art. Silke Ackermann concludes this section by charting the rapidly changing role of instruments in displays at the British Museum, probably best known for its focus on cultural history, over the course of 260 years.

But how do 'old' instruments and potential audiences keen to visit them fare in working institutions, whose main duty is night-sky observation rather than daytime displays? To give one (of many possible) answers to this question Laurence Bobis and Suzanne Débarbat chart how these challenges were met at the Paris Observatory from its founding in the late 1660s to almost the present day.

That instruments already took their place amongst many other products on display at the first "Universal exhibition" of 1851 in London has been studied in detail. However, Richard L. Kremer takes a different viewpoint, namely that of various groups who 'saw' those instruments on display at the first world's fair held in America, the Philadelphia Centennial Exhibition of 1876. Ten million visitors, more than ever before anywhere, saw scientific instruments offered by hundreds of exhibitors. What do their reports tell us about the cultural meaning of science?

The United States are also the focus of Steven Turner's and Richard A. Paselk's papers, highlighting yet another arena of display that poses quite distinct challenges: that of university museums, exemplified by the Science Teaching Museum at the University of Chicago (Turner) and the Robert A.

Paselk Scientific Instrument Museum at Humboldt State University (Paselk). How to put the university at the centre of the museum and the museum at the centre of the university? And how to manage displays and public as well as academic/teaching expectations when the museum is basically a private undertaking and a one-man-show as is the case at Humboldt State University?

Whilst the preceding papers all discuss different institutional challenges as well as changing attitudes that reflect a dynamic evolution of the institutional identity, Peggy Aldrich Kidwell and Amy Ackerberg-Hastings boldly take on an instrument that was for most of its life a work tool rather than an esthetically pleasing object designed for show as much as use: the humble slide-rule, displayed in the United States between 1840 and 2010. Why do these objects (or specific object groups in general, one might ask) merit our attention?

On quite a different stage (literally!) moves Ingrid Jendrzejewski with her discussion of how seventeenth-century English theatre responded to Galileo and the telescope following the publication of the *Sidereus Nuncius* in 1610. A different set is explored by Ileana Chinnici, Donatella Randazzo and Fausto Casi who investigate the role of (original) instruments on film-sets, focussing mainly, but not exclusively, on Visconti's *Il Gattopardo*, produced in the early 1960s.

Befitting for a book on instrument display, Inga Elmqvist-Söderlund rounds off the discussion by exploring the role of instruments displayed in books or more precisely in seventeenth-century frontispieces. The frontispieces and illustrated title pages contributed to shaping the identity of astronomy by suggesting that astronomical instruments were an appropriate attribute for a person who wanted to be associated with the field.

This volume includes only a dozen papers, a small selection of those presented at the 2010 conference; but nevertheless a colourful, multifaceted image emerges that shows that there can be no such thing as 'THE display', the quintessential answer to every prayer. Rather, displays of instruments are shaped by many different considerations that, nonetheless, all have one thing in common: the attempt to respond to constantly changing demands posed by the objects and their 'users', by the public and academic audiences, and by the evolving cultural, social and economic contexts in which the instruments, over time, have stood.

Silke Ackermann
Richard L. Kremer
Mara Miniati

Colour Plates

PLATE 1 *The globe room located on the first floor of the Palazzo Castellani, in a photograph taken in 1948* (PHOTOGRAPHIC LABORATORY AND ARCHIVE, MUSEO GALILEO, FLORENCE).

PLATE 2 *The 1964 Chemistry Galleries featured a reconstruction of the Government Chemist's Laboratory of 1895*
(© SCIENCE MUSEUM / SCIENCE & SOCIETY PICTURE LIBRARY).

PLATE 3 *Mechanical globe, by Georg Roll and Johannes Reinhold, 1584* (V&A 246-1865; © VICTORIA AND ALBERT MUSEUM, LONDON).

PLATE 4 *Astrolabes on display in the Jameel Gallery of Islamic Middle Eastern Art, 2006* (© VICTORIA AND ALBERT MUSEUM, LONDON).

PLATE 5 *Display "Transfer of knowledge between the Islamic World and Europe" in Room 40 at the British Museum* (© TRUSTEES OF THE BRITISH MUSEUM).

PLATE 6 *Opening of the exhibition*, François Arago et l'Observatoire de Paris, *22 September 2003.*

PLATE 7 *Keuffel & Esser's Philadelphia exhibit*
(PRINT AND PICTURE COLLECTION, FREE LIBRARY OF PHILADELPHIA, CEDC 021338. COURTESY OF FLP AND BRIDGEMAN ART LIBRARY).

PLATE 8 *Robert A. Paselk Scientific Instrument Museum located in Humboldt State University Library: main display cases, Left-hand (South) case.*

PLATE 9 *Fuller's time telegraph, 1860s* (SMITHSONIAN IMAGE 87-4843).

PLATE 10 *Patent model for John W. Nystrom's circular slide rule, 1851* (SMITHSONIAN IMAGE 87-8042).

PLATE 11 *Merz refractor belonged to the Prince Tomasi di Lampedusa* (INAF—OSSERVATORIO ASTRONOMICO DI PALERMO, MUSEO DELLA SPECOLA).

PLATE 12 *Late 19th-century cylinder phonograph, or 'parlograph', used in* The Best Offer *to record on a wax cylinder and reproduce the automaton's voice* (FAUSTO CASI COLLECTION, AREZZO).

PLATE 13 *Georg Welsch,* Commentarius in Ruznameh Nauruz . . ., *Augsburg, 1676: frontispiece by Johann Heinrich Schoenfeld and Johann Georg Waldreich*
(LIBRARY OF THE ROYAL SWEDISH ACADEMY OF SCIENCES AT STOCKHOLM UNIVERSITY LIBRARY. PHOTOGRAPH: STOCKHOLM UNIVERSITY LIBRARY).

CHAPTER 1

Andrea Corsini and the Creation of the Museum of the History of Science in Florence (1930–1961)

*Marco Beretta**

Introduction

The creation of the Institute and Museum of History of Science in Florence was due to the combination of a series of contingent circumstances driven by the strong-willed strategy of one man: Andrea Corsini (1875–1961) (*Fig. 1.1*). As surprising as it might seem, this curious combination is in fact quite typical in the history of Italian cultural heritage. Those passionate about the national heritage have rarely followed a deliberate and long-term political strategy and they have rarely been able to find support and funding from local or national authorities. The rescue and highlighting of invaluable collections, like the one put together by Corsini between 1925 and 1961, are almost miraculous and thus one might expect them to be short-lasting. The case of the collections of the Florentine museum fortunately tells a different story; these collections have seen three generations of curators and directors providing continuity, growth

* I wish to dedicate the present study to the memory of Daniela Carrara, late archivist of the Museo Galileo. The study, which was undertaken as part of the preparations for the 80th anniversary of the birth of the Istituto e Museo di Storia della Scienza, stems from research on manuscript and iconographic material from various sources, but principally the museum's archives. Some of the findings have already been published on the website of the museum, in the virtual exhibition Marco Beretta (ed.), *Museo Galileo 1930–2010*, 2010, http://www.museogalileo.it/istituto/en/index.html (accessed 19 March 2014). An opportune moment to undertake this research arrived with the decision to rearrange the historical archives of the museum and the correspondence of Corsini. I would like to thank Silke Ackermann, Francesco Barreca, Daniela Carrara†, Richard L. Kremer, Mara Miniati, Peter Morris, Tom Settle and Giorgio Strano for furnishing me with important information and many helpful suggestions. I thank Susan Meschel of the Museum of Science and Industry of Chicago for information and photographs relating to the models presented at the International Exposition *The Century of Progress* held in Chicago in 1933. Special thanks are extended to Carlo Maccagni for having provided me with valuable data on Pietro Pagnini, on the history of the museum during the 1950s and 1960s, and on the history of the sciences in Italy generally during this period.

 | DOI 10.1163/9789004264403_002

FIGURE 1.1 *Photograph of Andrea Corsini, taken in the early 1950s* (PHOTOGRAPHIC LABORATORY AND ARCHIVE, MUSEO GALILEO, FLORENCE).

and expansion. The present essay explores the early history of the Institute and Museum of History of Science and its changing approaches towards the exhibition of scientific instruments and collections.

Andrea Corsini

After receiving his degree in medicine from the University of Florence in 1899, Andrea Corsini became the assistant in 1902 of Giorgio Roster at the city's Ufficio d'Igiene. He remained at the Department of Public Hygiene for his entire career, becoming vice-director in 1906. After earning first place in a national competitive examination in 1928, he was promoted to the level of medical director, the highest grade in the municipal administration, fulfilling his duties in this position until his retirement in 1953.[1] Alongside his medical

1 Maria Luisa Bonelli, "È morto Andrea Corsini", *La Nazione*, 25 June 1961; Lino Agrifoglio, "Ricordo di Andrea Corsini", *Castalia* 17, 3 (1961), pp. 1–3; "In memoria di Andrea Corsini nel primo anniversario della sua morte", *Rivista di storia della medicina* 6 (1962), pp. 3–10; Vincenzo Cappelletti, Federico Di Trocchio, "Andrea Corsini", in *Dizionario biografico degli Italiani*, Rome, 1960–: vol. 29 (1983), pp. 606–609; Luigi Belloni, "Commemorazione di Andrea Corsini", *Nuncius* 4 (1989), pp. 177–190. On Corsini and the origins of the Institute and

activities, in 1909 Corsini was appointed lecturer on hygiene and public health at the University of Florence, a teaching post to which he applied himself diligently, publishing over the years eighty scientific papers on different topics in the areas of bacteriology and hygiene. However his true passion, to which he dedicated his life, laid in a completely different area.

It is not clear what first sparked Corsini's interest in the history of medicine, a subject on which he became a lecturer at the University of Siena in 1913. As a university student, he was one of the founding members of *Pro-Cultura*, a politically-inspired association with openly democratic ideals.[2] One of the objectives of *Pro-Cultura* was to promote the historical and cultural legacy of Florence. These activities set Corsini down a path in which he would become an early protagonist for the study of the history of science in Florence.

The First Steps towards a Museum of the History of Science

The determination and tenacity with which Corsini, from the opening years of the twentieth century, sought to conserve the scientific-historical heritage of his city and his country can be seen in a series of initiatives whose importance cannot be overestimated: the creation of the *Gruppo per la tutela del patrimonio scientifico nazionale* (Association for the Protection of the Scientific Heritage of Italy) in 1923; the founding of the Istituto di Storia delle Scienze (Institute for the History of Sciences) in 1925; the organization of the highly successful *Esposizione Nazionale di Storia della Scienza* (National Exposition on the History of Science) in 1929; and finally, on the 18th of May 1930, the inauguration of the Museo di Storia della Scienza (Museum of the History of Science) in Piazza dei Giudici in the heart of Florence. Behind these four achievements lay years of preparatory work that testify to Corsini's ability to marshal the most diverse array of individuals and circumstances in his support, channeling resources into the realization of projects which he pursued with scrupulous care and immense passion.

Museum of the History of Science, see Giovanna Baroncelli, Massimo Bucciantini, "Per una storia delle istituzioni storico-scientifiche in Italia: L'Istituto e Museo di Storia della Scienza di Firenze", *Nuncius* 5 (1990), pp. 5–52.

2 Regarding the founding of the association, Pietro Franceschini, a historian of medicine and a friend of Corsini, wrote: "Corsini was motivated by the political situation in Italy during the decade 1890–1898, already extremely dark in itself and promising little hope for the future". Franceschini to Luigi Belloni, 7 July 1962; Florence, Museo Galileo: Istituto e Museo di Storia della Scienza, Archive, (hereinafter referred to as "Fondo IMSS"), not yet catalogued.

The Istituto di Storia delle Scienze, inaugurated on 7 May 1925, occupied a modest pair of rooms in the library of the university's Faculty of Medicine but already prefigured the elements that would come together five years later in the Museo di Storia della Scienza. The objectives of this institute—which was destined, in Corsini's words, to become a "shrine to science", and in particular to the achievements of Italian scientists—were:

1. To create a library of the history of the sciences, divided into the areas of biology, physics, and mathematics;
2. To assemble a collection of iconographic material pertaining to scientists and their work;
3. To gather original artifacts documenting the history of the sciences, seeking above all to prevent its dispersal and to guarantee its conservation wherever it was located;
4. To collect any documentary material that could contribute to scholarly research on the history of science (autographs, photographs, prints, etc.).[3]

In the initial phase of this ambitious project, the University of Florence nominated a commission made up of prominent citizens—the businessman and politician Prince Piero Ginori Conti (1865–1939); the director of the Uffizi, Giovanni Poggi (1880–1961); and the distinguished astronomer Giorgio Abetti (1882–1982)—whose mandate was to "collect contributions from both amateur enthusiasts and academic scholars interested in the history of science" that would help to support the day-to-day running and "the further expansion of this most useful institution".[4] In addition, the commission would "discuss with the editorial board of the *Enciclopedia Italiana Treccani* and the *Dizionario biografico degli italiani* the possibility that the work already completed by the Istituto might be recognized and incorporated into the *Dizionario*, and that it [might be] invited to participate in this compilation".[5]

The institute, which overlooked the cloisters of the historic Ospedale di Santa Maria Nuova, immediately received significant donations of scientific

3 Fondo IMSS, *Carteggio. Primi anni di vita del Museo*, Istituto e Museo di Storia delle Scienze, 1926, f. 1. All the translations of the original Italian quotations included in this essay have been made by Lisa Chien.

4 Ibid., f. 2.

5 Ibid., f. 3. The museum archives preserve considerable material that had clearly been gathered for a projected illustrated encyclopaedia of Italian scientists, but no documentary evidence of a collaboration between the museum and the Istituto Treccani has yet emerged.

instruments and other exhibits, some of which still form part of the collection today. Noteworthy among these were: surgical instruments used by the eminent physician Giovanni Alessandro Brambilla for his obstetric operations;[6] some large bronze mortars produced by the foundry of Alessandro Tognozzi in 1764;[7] an exceptional collection of microscopes, including one designed by the scientist Giovan Battista Amici; anatomical specimens prepared by the polymath Girolamo Segato;[8] two *matule* (glass receptacles for collecting urine samples);[9] and a scale for weighing newborn babies.[10] To these were added, some time before 1927, a series of microscopic slides prepared by the anatomist Filippo Pacini;[11] an unspecified number of bottles containing chemical substances; two fine seventeenth-century alambics made of Murano glass; 18 plaster busts, 4 oil paintings, 20 framed portraits, 500 prints, 220 printed books, 125 manuscripts and other autograph material, and 800 items of miscellanea.[12]

As emerges even from this abbreviated inventory, when the institute opened it already vaunted a first-rate collection with some funding whose further expansion, as we will see, would only be hampered by a lack of space.[13] At the time, as a newspaper article signed by E. Lazzareschi and published in *La Nazione* on 5 February 1926 announced, Corsini was already proposing to enlarge the premises and to add "the last precious relics of the *Accademia del Cimento*" (a scientific academy founded in Florence in 1657), an indication that plans to move these historic exhibits from the *Tribuna di Galileo* at Palazzo Torrigiani, Florence, to the new museum had already been delineated.

The Istituto di Storia delle Scienze, the very first of its kind in Italy, felicitously combined two objectives: promoting public interest in the history of

6 Museo Galileo, Florence: Dep. Ospedale Santa Maria Nuova, Florence.

7 Museo Galileo, Florence: inv. no. 3600.

8 Girolamo Segato (1792–1836) was a naturalist, anatomist, Egyptologist and cartographer. Since 2001, these fragile exhibits are on loan at the University of Florence, Dipartimento di Anatomia, Istologia e Medicina Legale.

9 Museo Galileo, Florence: inv. no. 1779. During the 1966 flood one of the two *matule* was damaged beyond repair.

10 Fondo IMSS, *Inventarii*, 1925–1927.

11 Museo Galieo, Florence, inv. nos. 3269 etc.

12 Fondo IMSS, *Inventari*, 1925–1927.

13 The balance sheet in December 1927 included 15,500.70 Italian *Lire* in revenues, of which 10,000 *Lire* had been donated by Ginori Conti to help finance the costs of chartering the institute as a non-profit organization. "Istituto di storia delle scienze della R. Università di Firenze. Relazione al Consiglio di Amministrazione", March 1928: Fondo IMSS, *Carteggio. Primi anni di vita del Museo.*

science and raising awareness of the importance of collecting and conserving the instruments, documents, iconography, and other material connected with the sciences in Italy. In 1927 the institute was transformed by royal decree into an *Ente morale* or non-profit organization. No longer administered or financed by the city, it benefitted from the generous patronage of its new president, Ginori Conti (1865–1939), who provided the funding that allowed the museum to grow and extend its sphere of influence. Corsini laid out the objectives of the institute in a talk delivered before the Florentine literary and scientific academy *La Colombaria* in February 1927, which was entitled "Florence and the History of Science".[14] In it he spoke of his aspiration to transform the Istituto di Storia delle Scienze into a hybrid institution that would combine some of the characteristics of the prestigious Deutsches Museum in Munich, whose mission was to conserve Germany's scientific heritage for the benefit of its citizens, and the Institut für Geschichte der Medizin in Leipzig which, under the leadership of Karl Sudhoff (one of Corsini's correspondents), won the full acceptance of the scholarly community for the history of science as an area of academic study. Nevertheless, Corsini was at pains to underline that any direct comparison between the institute created by him and those already in existence (most of which focused on the history of medicine), such as the Museo Storico dell'Arte Sanitaria in Rome, were inappropriate, because the Istituto di Storia delle Scienze was connected with the University and, as such, its primary purpose was to contribute to scholarly research.[15] The specific nature and characteristics of the institute were summarized in a circular published in 1927:

> The Royal University of Florence has [established] an 'Institute of the History of the Sciences' that, being dedicated to this special branch of research, *not only attends to the bibliographic and iconographic aspects, but collects and arranges the historic scientific material in such a way as to create a unique and most important museum in central Italy for these studies.*[16]

14 Andrea Corsini, "Firenze e la Storia delle scienze", *Atti della Società Colombaria* (1926–1927), pp. 3–9: 8.

15 Corsini prepared these remarks, which were published in the 15 November 1925 issue of *Medicina nuova*, to allay the fears expressed by the editors of the medical journal that the museum in Florence might set itself up as a rival to Rome's museum of the history of medicine.

16 The circular was signed by Piero Ginori Conti, Giovanni Poggi and Giorgio Abetti, members of the commission nominated by the University of Florence to help manage the

FIGURE 1.2 *The main hall of the* Tribuna di Galileo (*1841*) *in the museum* La Specola, *shown in a late-19th century photograph*
(PHOTOGRAPHIC LABORATORY AND ARCHIVE, MUSEO GALILEO, FLORENCE).

Overcoming Obstacles

In pursuing his dream of creating a museum dedicated to the history of science, Corsini came up against vested interests connected with a well-established and historic institution, the Museo degli Strumenti Antichi (Museum of Antique Instruments), which was housed in the *Tribuna di Galileo*—a magnificent set of rooms in a palace that the Grand Duke of Tuscany Leopold II (1797–1870) had decorated with frescoes recounting the life and achievements of Galileo Galilei (*Fig. 1.2*). Earlier Andrea Corsini and the *Gruppo per la tutela del patrimonio scientifico nazionale* led by Antonio Garbasso (1871–1933) had combined their efforts to protect the future of the *Tribuna* and its collections, and succeeded in drawing up an agreement with the University of Florence that was signed on 4 March 1925, in which they were formally charged with

finances of the museum. Printed circular, Fondo IMSS, *Carteggio. Primi anni di vita del Museo*, Italics mine.

re-organizing the collections and re-opening the museum to the general public.[17] Due to the lack of funds, they found themselves not at liberty to take any concrete action, and Garbasso, who was also the director of *La Specola* (the old observatory), realized that in order to save the collections it would be necessary to adopt another strategy. For this reason, shortly afterwards he took the unexpected step of having the museum's valuable collection of books and manuscripts moved to the astronomical institute and new observatory at Arcetri, in the vicinity where Galileo spent his last years under house arrest. Garbasso's plan was to later move the museum collections there as well.

Acutely aware himself of the historic importance of *La Specola*'s collections, Corsini set in train his own plans as an alternative to those of Garbasso. A period of intense activity followed in which Corsini, assisted by Garbasso and Ginori Conti, sought to raise public support for their initiative of creating a museum of the history of science. One of their most successful ventures was to organize the first *Esposizione Nazionale di Storia della Scienza*. Inaugurated on 8 May 1929, this exhibition was received with such enthusiasm that it seemed to bring palpably closer to realization Corsini's dream of establishing a museum in Florence dedicated to the history of science.

More than 100,000 visitors filed through the Florentine pavilion, whose exhibits documented the extent and richness of the Tuscan capital's scientific heritage.[18] It also served to underline the lamentable fact that this patrimony was dispersed in various dusty university departments, obscure and inaccessible historic buildings, and museums that had been half-closed for years,[19]

17 "Gruppo per la tutela del patrimonio scientifico nazionale. Comunicazioni ufficiali", *Archivio di storia della scienza* 6 (1925), p. 95.

18 The importance of this legacy was exhaustively described in the volume prepared by Giuseppe Boffito (1869–1944) for the exhibition, and which he dedicated to Ginori Conti: *Gli strumenti della scienza e la scienza degli strumenti*, Florence, 1929. It consisted of an expanded edition of the catalogue compiled at the end of the nineteenth century by Ferdinando Meucci, providing a historical context for the collection and filling in many lacunae. In compiling *Gli strumenti* Boffito benefitted from the collaboration of Giulio Cipriani, who was the curator, under the directorship of Garbasso, of the collection of scientific instruments in the museum *La Specola* before becoming curator of the collections of the new Museo di Storia della Scienza directed by Corsini.

19 On display in the Florentine pavilion of this national exposition were 303 scientific instruments from the Museo degli Strumenti Antichi; 68 manuscripts, books and portraits from the library of the University of Florence's Faculty of Medicine; 59 items, including chemical samples and laboratory instruments, from the chemical laboratory of the University of Florence; 77 items ranging from manuscripts and zoological specimens to portraits and busts from the Laboratorio di Zoologia, the Laboratorio di Anatomia Comparata and

a situation crying out for a fresh initiative that would make possible the appreciation of the city's heritage. The exposition was so well planned and realized that it received the unconditional praise of Prince Ginori Conti. In an address delivered to an audience that included King Vittorio Emanuele III, the prince underlined the primary goals of the exhibition: 1. the publication of a catalogue that would document not only the scientific exhibits on display but also, in more than 3,000 short biographies, the many significant contributions of Italians to the sciences; 2. to rekindle "not only in the latest generation of scientists, but also in the youth of Italy a love for the history of science"; and, finally, 3. the creation of a "Museo Italiano di Storia della Scienza" in Florence, which would house "the first large inventory of historical-scientific material in Italy's possession".[20] These were the same objectives that had motivated Corsini to found the Istituto di Storia delle Scienze in 1925.

The success of the *Esposizione Nazionale*, however, awakened the jealousy of some of Corsini's colleagues who sought, by launching their own initiatives, to gain credit for the realization of the first science history museum in Italy. In September 1929, with the exhibition still open and attracting large numbers of visitors, the physicist Carlo Del Lungo, son of the more famous Isidoro Del Lungo (1841–1927) who had edited the works of Galileo, undertook to persuade the Ministry of Culture to establish a "Museo storico della scienza Italiana" that would be completely independent from the University of Florence. His proposal, which had first been advanced by Corsini in 1923, was to "re-organize

---

the Museo Zoologico; 48 books and mineral samples from the Istituto di Mineralogia; 208 samples and a portrait of Bellini from the Istituto di Farmacologia; 15 documents, portraits, and models from the Istituto di Antropologia, Etnologia e Paleontologia; 29 herbals, wax models and models made from other materials lent by the Istituto Botanico; 32 anatomical exhibits and 32 books and instruments from the Istituto Anatomico; and 26 instruments, documents and portraits from the Osservatorio Astronomico in Arcetri. The original documents detailing the loans are conserved in the *Archivio Generale* of the University of Florence and the museum has a copy of these documents: Fondo IMSS, *Raccolte manoscritti* 32. The exhibition was particularly comprehensive, because it included not only objects borrowed from various departments at the university, but also many items loaned by private individuals. Regarding the additions to the collection of the Museo di Storia della Scienza after the exposition, we may mention the room of exhibits donated by the Pelli-Fabbroni family in memory of the Fabbroni brothers; the family would later cede various important documents to Corsini.

20 "Sulle origini, costituzione e sviluppo del Museo Nazionale di Storia delle Scienze", 1933: Fondo IMSS, *Carteggio. Primi anni di vita del Museo*. This is a 12-page typewritten document, probably prepared by Corsini, with ample passages drawn from the address given by Piero Ginori Conti.

and revive as an effective and worthy institution our historic R[egio] Museo di Fisica e Storia Naturale, set up in the Tribuna galileiana, restoring its original name and making of it once again the great Museum of the History of the Sciences, natural history and medicine, as it deserves".[21] The true motives behind this apparently disinterested proposal emerge in a letter that Del Lungo sent to the Direzione Generale delle Antichità e Belle Arti of the Ministry of Culture. In it he urged the director to nominate as the head of the new institution "a professor of physics at the university or secondary school level, from among those in possession of some expertise in the History of the Sciences".[22] He was patently recommending himself as the most suitable candidate, seeking to exclude Corsini with his position in the much less prestigious municipal Department of Hygiene.

The interest generated by the 1929 exhibition and the marks of recognition that were conferred on the Istituto di Storia delle Scienze, as well as the unflagging support of Ginori Conti and Garbasso, encouraged Corsini to counter the machinations of Del Lungo by accepting the suggestion of Angelo Bruschi, director of the Biblioteca Nazionale Centrale of Florence, to move his institute and nascent Museo di Storia della Scienza into the Palazzo Castellani in Piazza dei Giudici, which the Biblioteca Nazionale was supposed to be vacating.[23] Perhaps his decision was somewhat hastily taken, but the unhappy fate of the collections of *La Specola*—threatened with exile in Arcetri's observatory—reinforced Corsini's conviction that he had to find premises compatible with his museum's innovative historiographic and cultural objectives, rather than assenting to the offer of the more prestigious premises of the *Tribuna*.

By reacting promptly, Corsini was able to circumvent the plans of Garbasso, already partially realized with the transfer of the library of the Museo degli Strumenti Antichi to Arcetri,[24] and the more pernicious strategems of Del Lungo. In 1930 he transferred his modest Istituto di Storia delle Scienze—

21 Del Lungo to Corsini, 26 September 1929 and 5 October 1929: Fondo IMSS, 1924–1977, *Carteggio e altri documenti ufficiali*. Fortunately for Corsini, Del Lungo's plans were strongly opposed by Garbasso.

22 Unsigned typewritten copy of a letter written by Del Lungo, 1929: Fondo IMSS, *Carteggio. Primi anni di vita del Museo*.

23 Minutes of a meeting of the board of governors of the museum held on 6 December 1930, Archivio Storico del Comune di Firenze (ASCFi), Florence, CF 9260.

24 An agreement drawn up between the University of Florence and the Istituto di Storia delle Scienze on 29 June 1933 states in its first paragraph: "that to the said Institute [should be] consigned all of those books of historical value, which form part of the Library of Antique Instruments of Physics and Astronomy and that are presently stored in the Laboratory of Physics". Fondo IMSS, 1924–1977, *Carteggio e documenti ufficiali*. These books came into

enriched by the addition of many items from the 1929 exhibition—to the new seat in Piazza dei Giudici that had been assigned to it by the Ministry for Public Education. However, the Palazzo Castellani had various longstanding occupants who were loath to leave: the Biblioteca Nazionale Centrale, which had a store room for its books on the ground floor and a reading room on the first floor; the Accademia della Crusca, a cultural and literary academy founded in Florence in 1570–1580, on the second floor; and the *Deputazione di Storia Patria* on the top floor. The entire palazzo might have been placed at the disposition of the Istituto and Museo immediately, but Corsini never had the chance to see this project completed. In 1929 only the recently restored rooms on the ground floor were made available to him.[25]

The Museo Nazionale di Storia della Scienza was inaugurated on 18 May 1930 at 4:30 pm by Benito Mussolini.[26] There are no known photographs or published accounts of this event, making it impossible for us to determine with any certainty the size and arrangement of the collections. However, a typewritten report probably prepared by Andrea Corsini (who had been appointed director of the new museum) shortly after its inauguration provides some details regarding the exhibition layout and is therefore worth quoting at length:

---

the museum's possession only after the war, however, in two separate consignments in 1948 and 1950.

25 The ground floor consisted of eleven rooms of varying size, but their layout was unsuitable for a museum as they could not be visited in uninterrupted sequence. The Accademia della Crusca moved to the Villa Medicea di Castello in 1972, and the Deputazione di Storia Patria finally transferred its premises to the Palazzo Neroni in 1996. See: Luigi Zangheri, "La sede del Museo Galileo", in Filippo Camerota (ed.), *Museo Galileo: Capolavori della scienza*, Florence, 2010, pp. 56–57. For a visual history of the museum, see Marco Beretta, "Imaging the Museum of History of Science (1841–1991)", in Camerota (ed.), *Displaying Scientific Instruments: From the Medici Wardrobe to the Museo Galileo*, Florence, 2012, pp. 151–200.

26 The name of the new institution was not formalized in an official decision. However, a letter dated 13 January 1931 and sent to Piero Ginori Conti from the Under-Secretary of State on behalf of the President of the Council of Ministers, Benito Mussolini, mentions the "National Museum of the History of Science". Fondo IMSS, *Carteggio. Primi anni di vita del Museo*. In fact, during its history the name of the museum, designed by the various directors, varied often and—it seems—somewhat haphazardly, e.g.: Istituto Nazionale di Storia della Scienza in 1924; Istituto di Storia delle Scienze in 1925–1927; Istituto e Museo di Storia delle Scienze in 1929; Museo Nazionale di Storia delle Scienze in 1930; Istituto e Museo di Storia delle Scienze in the 1930s; Istituto e Museo di Storia della Scienza from 1945 onward (with an occasional return to Museo di Storia delle Scienze); and finally, in 2010, Museo Galileo: Istituto e Museo di Storia della Scienza.

In May 1930 the Museum of the History of Science, housed in Palazzo Castellani on Piazza dei Giudici, was solemnly inaugurated by the Head of State.

In this Florentine Museum is a collection that may be regarded as complete, to the point of forming in itself a compendium of international interest.

It includes astronomical, geodesic, and gnomonic instruments from the sixteenth and seventeenth centuries, notably E. Danti's primum mobile (1578), the quadrant of Cosimo II (1609–1627), Giusti's quadrant of 1563, various drawing compasses, plumb levels, and cross-staffs. Torricelli's original barometer built in 1644; instruments by Vincenzo Viviani; two of his globes, one celestial, the other terrestrial, one built in 1632, the other in 1636.

The room of Galilean relics contains his first two telescopes with magnifications of 14 and 20, made in 1609; a finely ground objective lens, a natural lodestone with an armature, a proportional or military compass from 1596, the diagram of an escapement mechanism for the application of the pendulum to clocks, a thermoscope and, in a small crystal urn, the index finger of the great philosopher's left hand. The same room also displays a very important series of telescopes made by Torricelli, Viviani, Mariani, Divini, and Campani, and a large series of magnifying glasses and objectives, including exemplars remarkable for their fine workmanship made by Torricelli, Divini, and Campani, the eyeglass of Pope Leon X, and other items.

In the adjoining rooms are precious instruments that were used by members of the *Accademia del Cimento*: the hygrometers of Vincenzo Viviani and Francesco Folli, and the condensation hygrometer of Grand Duke Ferdinando II, clinical condensation thermometers, hydrometers, and copper and tin spheres employed by the Academicians for experiments on the compressibility of water; in addition, in an adjacent room, the chemistry cabinet of Grand Duke Peter Leopold with ivory mortars, agate mortars, quartz vases, etc.

Worthy of mention is the very large collection of astrolabes, perhaps the world's most important, including an astrolabe made in 1102–1103 (year 495 of the Hegira), Arscenius's astrolabe, Duddley's [sic.] astrolabe, Barman's [sic.] astrolabe, etc. There is also the large burning-lens, 0.41 [meters] in diameter, with which H. Davy and his assistant Faraday performed their experiment on the combustion of diamonds and other precious stones.

In other rooms there are anatomical parts petrified by G. Segato (1792–1836) and the large Viennese armamentarium donated by Grand Duke Peter Leopold in 1795, consisting of 30 drawers of surgical instruments. Nearby is a display case devoted to Dr. L. Gigli, inventor of the wire saw.

And in addition sundials, sidereal dials, of different shapes and sizes and also from the Galilean period, an original of Leopoldo Nobili's first galvanometer with astatic needles, and a series of thermo-electric piles made by the same hand. One can furthermore admire a fine series of microscopes by Patroni and Pacini, and dioptric and catadiopotric microscopes by Prof. Amici from 1812.

Such an assemblage constitutes one of Italy's most important collections, and the public should acquaint itself with this new museum, which offers further evidence of all the fields so superbly addressed by Italian genius.[27]

Another profile of the museum, written on the occasion of its inauguration by Luigi Castaldi (1890–1945), a friend of Corsini and an authoritative scholar on the history of medicine, appeared in *Archeion*, the journal of the history of science founded by science historian Aldo Mieli (1879–1950). It is worth citing an extended passage from this article, for it paints a vivid picture of the museum in its early days:

> Today the Museum occupies the ground floor with 9 rooms for exhibits, an administrative office and a library. All 9 rooms are already filled with extremely interesting items documenting the past, some of which are truly unique and precious. Galileo's famous telescope, the telescopes of Torricelli, Giuseppe Campani, Eustachio Divini; many microscopes including some constructed by Amici, and the Lilliputian one depicted by me [...]; some very valuable maps of the world; Nobili's double magnet [...] and many physics instruments that once belonged to the *Gabinetto degli Strumenti antichi* of *La Specola*. In addition, I may cite the Chinese lever scales, antique telescopes, early versions of the internal combustion engine, etc. The walls are decorated with paintings and busts, and the collection of medals awarded to the museum at the *Congressi italiani degli scienziati*. Of great interest are models of Leonardo's flying machines; replicas of grinding mills from Pompeii; irrigation devices used

27 Corsini [?], Two typescript pages with corrections, 1930; Fondo IMSS, *Carteggio. Primi anni di vita del Museo.*

> in the region of Lucca; murals depicting Sardinian *nuraghi* [Bronze Age tower constructions], a Roman sewer system, an homage to Alessandro Volta, and the sun tower of Arcetri; etc.[28]

At its opening, therefore, the museum consisted of nine rooms, all situated on the ground floor of the Palazzo Castellani,[29] and a collection that was already of outstanding quality even though at the time it only included a small part of the holdings of the *Gabinetto degli Strumenti antichi* (Cabinet of Antique Instruments), items that had been exhibited at the 1929 *Esposizione Nazionale*. However, the failure of the other occupants to vacate the upper floors of the building as promised made it impossible for the museum to expand its collections or draw up plans for the permanent installation of its exhibits. In addition to the instruments listed above, the 1929 exposition left the museum a legacy of assorted furnishings, display cases, and iconographic and printed materials which obliged the museum to use a part of its limited space for storage.

The planned move of the Biblioteca Nazionale advanced with frustrating slowness, and the museum was not able to open its doors to the public until late April 1931.[30] When it finally did, its layout had been altered and improved from the time of Mussolini's visit eleven months earlier. An article by Piero Domenichelli that appeared in *Il Popolo d'Italia* on 31 December 1932, testifies to the significant progress made, describing the exhibits in seven rooms:

> To the left of the entrance was Room 1. Astronomical, geodesic, and gnomonic instruments. Torricelli's barometer and Viviani's instruments.

---

28 Luigi Castaldi, "Il Museo Fiorentino di Storia delle scienze", *Archeion* 13 (1931), pp. 66–67. The ancient Roman artefacts, irrigation devices, and four wall murals were ceded, probably in the aftermath of the 1966 flood although the archives are silent regarding their destination. It is not clear to what Castaldi was referring when he cited the "Volta's memorabilia", and in any event their present whereabouts are unknown.

29 A 1927 plan of the ground floor of the Palazzo Castellani conserved in the ASCFi, Florence, shows that there were actually ten serviceable rooms. The rooms not used to display the collections probably housed the library and administrative offices or served as storerooms.

30 In February 1931 the Municipality of Florence approved an annual subsidy for the museum of 8,000 *lire* (roughly 8,000 Euros in today's currency): "Deliberazione del Podestà", 17 February 1931: Fondo IMSS, 1924–1977, *Carteggio e altri documenti ufficiali*. This subsidy, which was instrumental to the museum in the preparations for its opening, does not seem to have been renewed in later years. On this point, see Ginori Conti to Corsini, 31 March 1931: Fondo IMSS, 1924–1977, *Carteggio e altri documenti ufficiali*.

FIGURE 1.3 *Photograph published in 1932 showing the room dedicated to the museum's collection of Galilean instruments. Situated between the Piazza dei Giudici and the Lungarno (the avenue running through the heart of Florence along the Arno River), this room today houses the Museo Galileo bookshop*
(PHOTOGRAPHIC LABORATORY AND ARCHIVE, MUSEO GALILEO, FLORENCE).

Room 2. Galilean relics.

Room 3. Torricelli's telescopes, binoculars by Cherubin d'Orleans, lenses, podometer, magnetic compasses, goniometer, plumb levels, Rinaldini's quadrant, and glass objects and hygrometers from the *Accademia del Cimento*.

Room 4. Peter Leopold's chemistry cabinet.

Room 5. Petrified anatomical models by Girolamo Segato and a surgical armamentarium donated by Peter Leopold in 1795.

Returning to the entrance and turning left, the visitor entered Room 6 devoted to air pumps, Hero's fountain, and a collection of sundials and nocturnal clocks.

Room 7. Nobili's instruments and Pacinotti's electro-magnetic ring.

The article was accompanied by a photograph of the Galileo Room showing two telescopes and other Galilean memorabilia, as well as an exemplar of Bregans's "burning-lens" in a fine wooden mount (*Fig. 1.3*).

The possibility of acquiring on permanent loan the collection of the Museo Copernicano (founded in Rome in 1882 by Arturo Wolynski) arose,[31] but the decision appeared to be blocked by logistic problems and in the summer of 1932 Corsini launched into intense negotiations to have the Roman museum's holdings transferred to Florence. His intention was to place them on display in the Palazzo Strozzi, although such a magnificent Renaissance palace was still in family hands at the time,[32] but Ginori Conti did not believe that it offered sufficient space for the collections.[33] In the end, this particular project had to be abandoned.

Nonetheless, the future of the museum did appear to be somewhat brighter, because it had successfully collaborated on a project conceived by Guglielmo Marconi that had won the full backing of Mussolini: the preparation of a pavilion dedicated to the sciences in Italy (featuring among its exhibits replicas of some of Marconi's own inventions) for the international exposition *The Century of Progress* which was to be held in Chicago in 1933–1934.[34] For the

31 On this affair, see the letter by Federico Millosevich to Corsini, 23 November 1929: Fondo IMSS, 1924–1977, *Carteggio e altri documenti ufficiali*. Not only was the collection never transferred from Rome to Florence, but Corsini feared the very real possibility that the Museo Copernicano might be transformed into the officially recognized Istituto e Museo Nazionale di Storia della Scienza. Almost simultaneously, in 1933 Guido Ucelli announced plans to create a Museo Nazionale della Scienza in Milan which, as events transpired, would not be inaugurated until after the war, in 1953. The Florentine intellectual Sebastiano Timpanaro weighed into the debate with an article entitled "Illuminazioni scientifiche. Firenze per la storia della scienza", which defended the autonomy and central importance of the museum in Florence. A cutting of this article, which was published on 2 March 1934, is conserved in Fondo IMSS, *Ritagli*, 39/58. Even if, as has often been underlined, the proposals to establish analogous museums in Rome and Milan were a source of anxiety for Corsini, because they posed a threat to the future of his museum in Florence, they also attest to the growing public and academic interest in the Italian scientific heritage.

32 The details regarding these negotiations are not known. However, in 1937 the family sold the Palazzo Strozzi to the Istituto Nazionale delle Assicurazioni.

33 Ginori Conti expressed his doubts in a letter to Corsini dated 5 July 1931: "Regarding the change in the museum, which is supposed to be transferred to the Palazzo Strozzi, we should reflect on it carefully, because the place is exceptionally beautiful, but [the museum] will require much space. Will there be? Will they give us what is necessary?" Fondo IMSS, 1924–1977, *Carteggio e altri documenti ufficiali*.

34 Here is the complete text of the letter sent by Marconi to Corsini on 29 October 1932: "Illustrious Professor / His Excellency the Head of the Government has decided that Italy should participate in the International Exposition in Chicago with a collection of copies of valuable relics and documents aimed at demonstrating the extent of the contribution

Fascist government this exposition constituted an event of strategic importance, while for Piero Ginori Conti and Andrea Corsini it offered the opportunity to demonstrate the primacy of their museum and its collections. The contribution of the Museo di Storia della Scienza to the project was in fact crucial both in the planning phase,[35] and in its realization.[36] A key role was

---

made by our country to the scientific and technical progress of humanity. This collection will then pass to the American Museum of the Sciences of Chicago. Taking advantage of the preparatory work necessary for this exhibition, His Excellency the Head of Government has decided that there should not be just one, but four collections prepared, of which one is destined to be conserved in Italy, while one is destined for the Museum of Sciences in London, and the other for the Museum of Sciences in Munich, Bavaria. / It would be desirable that this collection include documents and relics connected with the most noteworthy contribution that Italians have made to progress in the medical sciences. / Also in relation to the agreements arrived at during the meetings held by the *Società per la storia della medicina* [Society for the History of Medicine] I would beg Your Excellency to be so good as to gather together all the documentation pertaining to Borelli regarding which you could certainly most authoritatively occupy yourself, given your well-known expertise in this area. / I hope that Your Excellency will be disposed to help us in this anything but simple task, which nonetheless represents our patriotic duty. [...] It will be necessary, however, for Your Excellency to bear in mind that in the most categorical way at least one copy [of this collection] must be ready by the 15th of February". Fondo IMSS, *Carteggio Corsini I.* We know in fact that Corsini did carry out research on Borelli, even if no publication came out of this work.

Corsini's hope of obtaining, in addition to official recognition of the Museo Nazionale di Storia della Scienza, a promise of regular financial support from the Consiglio Nazionale delle Ricerche (CNR) were destined to be disappointed.

35 In the letter written to Marconi on 3 November 1932, Ginori Conti proposed that replicas of 28 of the most important exhibits in the museum's collection be sent the Chicago exposition. In addition to the Galilean instruments, the prince suggested sending a replica of Santucci's magnificent armillary sphere, which was still conserved in the museum *La Specola*. His list included replicas of the telescopes and other instruments made by Galileo, Rinaldini's quadrant, various instruments from the *Accademia del Cimento*, the barometer and telescopes devised by Torricelli, the armillary spheres made by Vincenzo Viviani and Girolamo Della Volpaia, and different pneumatic pumps and instruments developed by Leopoldo Nobili. Next to each item was a note indicating the estimated cost of building a replica. It was anticipated, for example, that a reconstruction of Santucci's sphere would come to 45,000 *lire* (ca. 45,000 Euros). A copy of the letter from Ginori to Marconi is conserved in Fondo IMSS, 1924–1977, *Carteggio e altri documenti ufficiali.*

36 The replicas that were actually made and sent to Chicago included the following: telescopes, a thermoscope and a verge escapement mechanism constructed by Galileo; a barometer by Torricelli; a hygrometer devised by Ferdinando II; an odometer made by the academicians of the *Cimento*; a thermometer made by the academicians of the *Cimento*; a helicoid thermometer and a medical thermometer from the collection of the

played by Giulio Cipriani, the extremely able curator of the museum; given the task of putting together the exhibit of historic instruments for the exposition, he took advantage of the tradition of Florentine craftsmanship to have many fine replicas made *ex novo*, and also to borrow back models that had been produced in preceding years and sent "to museums in various part of the world, from England to Australia".[37] The exhibit was a resounding success with visitors to the exposition in Chicago, but did not lead to the hoped-for institutional support in Italy, either locally or from Rome, and the museum continued to struggle on a limited budget.

All the same there were positive developments. In 1933 the situation at the Palazzo Castellani began to resolve itself, with the Biblioteca Nazionale ceding its rooms on the first floor to the museum.[38] With this expansion of its floor

---

Accademia del Cimento; Nobili's galvanometer; a telescope lens by Torricelli; a model of Bregans's double lens; a replica of the apparatus devised by Macedonio Melloni to study infrared rays; Giovanni Amici's direct vision prism; a replica of one of Amici's microscope objectives; a replica of Eugenio Barsanti and Felice Matteucci's prototype for the first internal combustion engine; and a replica of Caselli's pantelegraph. The list, which appears in the letter of reply from Marconi to Ginori Conti dated 21 January 1933, did not correspond exactly to the number of items in the list sent in November by Ginori Conti in Florence to Marconi. Bowing to the Fascist agenda of promoting the applied sciences, some 'new' inventions were added such as the Barsanti-Matteucci engine and Caselli's pantelegraph. Some of the original replicas sent by the Istituto di Storia della Scienza to Chicago for the exposition are still conserved today in the storerooms of the Museum of Science and Industry in Chicago.

37 Letter by Ginori Conti to Marconi, 3 November 1932, cit. (n. 35). From the archives we know that between 1923 and 1930 Cipriani had the following replicas prepared for the Science Museum of London: telescopes by Galileo (inv. nos. 1923–667 and 1923–668); the hygrometer devised by Ferdinando II (inv. no. 1930–638); Galileo's thermoscope (inv. no. 1930–650); Torricelli's barometer (inv. no. 1930–638); a thermometer made by the grand duke's glassblowers (who were nicknamed the *Gonfia*—the 'puffed up' ones) (inv. no. 1930–651); and a second thermometer with an unusual heliocoid shape, made by the *Gonfia* (inv. no. 1930–652).

38 "This Museum of the Royal University of Florence has been enlarged and better arranged in the Palazzo dei Giudici"; *Rivista di storia delle scienze mediche e naturali* 15 (1933), p. 96. The Biblioteca Nazionale would continue to occupy the second floor of the Palazzo Castellani for some decades. The museum even contributed 1,000 *lire* to the costs of moving a part of the library's collection into a storehouse in the neighbourhood of Santa Croce. On 13 June 1934 Ginori Conti wrote yet again to the director of the Biblioteca Nazionale requesting that he vacate the second floor of the palazzo. He received a response two days later in which it was promised that the Biblioteca Nazionale would remove from "the upper floors of the Palazzo dei Giudici [...] the library furnishings" by the summer of the following year. Fondo IMSS, 1924–1977, *Carteggio e altri documenti ufficiali*.

FIGURE 1.4 *Palazzo Castellani, seat of the Museo di Storia della Scienza, in a photo taken during the early 1930s*
(PHOTOGRAPHIC LABORATORY AND ARCHIVE, MUSEO GALILEO, FLORENCE).

space, an agreement was signed with the University of Florence regarding the criteria for the conservation of the museum's collections.[39] Even though its staff during this period was limited to two persons (the curator, Giulio Cipriani, and the librarian, Rosa Franceschini), the museum managed by the end of the year to set up its first exhibition. Once again the lack of photographic documentation makes it difficult to visualize the developments taking place during what was a most significant phase in the history of the museum (*Fig. 1.4*).

From an enthusiastic, but all too brief description of the new layout published by Giorgio Abetti in February 1934,[40] we learn that the museum had doubled the number of its rooms to fourteen and was able to spread the display of its collections over two floors. The first room contained astronomical,

39 "Convenzione fra la R. Università di Firenze e l'Istituto di Storia delle Scienze", 29 June 1933 (four typewritten pages divided into 14 articles): Fondo IMSS, *Carteggio. Primi anni di vita del Museo.*

40 Giorgio Abetti, "Le origini e lo sviluppo del Museo di Storia delle Scienze in Firenze", *La scuola superiore* 1, 11 (1934), pp. 588–598.

geometrical, and gnomonic instruments dating to the sixteenth and seventeenth centuries, Torricelli's barometer, and Viviani's instruments. The second room was devoted to the scientific achievements of Galileo, complemented by the addition of telescopes from the Galilean school (previously exhibited in another room) and a celestial globe of Arab manufacture.[41] The documentation available, in particular Corsini's correspondence, provides us with some idea of the contents of the eight other rooms: Room III, Girolamo Della Volpaia's armillary sphere and Benedetto Bregans's double lens mounted in a wooden stand made by Francesco Spighi; Room IV, the collection of the *Accademia del Cimento*; Room V, a reconstruction of Peter Leopold's chemistry cabinet; Room VI, exhibits pertaining to the history of medicine; Room VI, Girolamo Segato's artifically petrified anatomical specimens; Room VIII, microscopes; Room IX, instruments and relics of Alessandro Volta; and Room X, electrical apparatuses. It may be presumed that, with this expansion and reorganization of the museum's exhibition space, Corsini finally felt he could take the decision to allocate a room for the institute's library and another for the director and his staff. In addition, we know that by the end of the 1930s the *Società Italiana di Storia delle Scienze Mediche e Naturali* (Italian Society for the History of the Medical and Natural Sciences) was regularly holding meetings on the first floor of the museum.[42]

Even though the number of visitors was relatively modest (*Table 1*),[43] it is interesting to note that in 1935 the museum had very generous visiting hours, being open to the public seven days a week, both in the morning and in the afternoon.[44] Many foreigners came, including a conspicuous number of

41 Abetti provided a description of these two rooms, but remained vague about the rest. Equally lacking in detail was the article "Il Museo Nazionale di Storia della Scienza", *L'Ambrosiano* (6 February 1934); Fondo IMSS, *Ritagli* 39/58, which did however include an interesting photograph documenting the change in the display of Galileo's telescopes. In later decades Galileo came to symbolize the museum. The arrangement of the telescope exhibit reproduced that of the 1929 *Esposizione Nazionale*, as may be seen in Giuseppe Boffito, *Gli strumenti della scienza e la scienza degli strumenti*, Florence, 1929, pl. 109.

42 Corsini became president of the association in 1940. It is not clear whether the display cabinets were regularly removed from the large hall on the first floor when these meetings were held or if, as would appear from photographs taken during the meeting held on 5 October 1941, the hall was no longer being used for the museum exhibits.

43 The number of visitors recorded in the table is approximate because it is based on the only source of information available—the signatures in the visitor's album; Fondo IMSS, not yet catalogued.

44 In a letter dated 25 April 1935 and sent to a publishing house in Germany in response to a query for information regarding the museum, Corsini wrote: "The Museum is open

TABLE 1 *Number of visitors who signed the Museum Guestbook per year.*

Year	Number of visitors
1931	399
1932	562
1933	474
1934	385
1935	321
1936	417
1937	303
1938	296
1939	194
1940	27 (as of 20 June 1940, after which the museum remained closed until June 1945)

science historians, a clear indication that the museum and the institute constituted the two inseparable halves of a single entity dedicated to education, research, and the conservation of Italy's cultural heritage.

A difficult interregnum followed, as the country sank into the darkness of an autocratic self-celebratory dictatorship. Corsini, however, assiduously cultivated ties with colleagues of the highest level around the world, and his correspondence with Arnold Carl Klebs, Max Neuburger, George Sarton, Henry Siegrist, Charles Singer, Karl Sudhoff, Erik Waller and other historians of science and medicine provides ample evidence of his desire to immunize the museum against the vacuous nationalist rhetoric promoted by the regime.[45] Indeed, although his resistance was discrete, Corsini was never a supporter of fascism. After the institution in 1938 of the racial laws that deprived Italian Jews of their citizenship and the right to hold positions in the professions or in government, Corsini was one of the few intellectuals to maintain close ties with the medical historian Arturo Castiglioni (1874–1953), who was forced to

every day and observes the following opening hours: Winter 9:00–12:00 and 14:00–17:00, Summer 9:00–12:00 and 15:00–18:00, Sundays 10:00–12:00. Entrance ticket 2 *lire* [about 2 Euros]". Fondo IMSS, *Carteggio. Primi anni di vita del Museo.*

45 From Corsini's correspondence with these scholars it is clear that he also devoted considerable attention to the expansion of the museum library's collection of books, an undertaking that would be continued by Maria Luisa Bonelli after his death.

emigrate to the United States in 1939 (where he became a professor at Yale University), and with Aldo Mieli (1879–1950), who sought refuge in Argentina. In a letter sent from Buenos Aires by the Argentinian industrialist and science historian Umberto Julio Paoli (1876–1953),[46] we learn that after Corsini and Luigi Castaldi[47] complained bitterly to him of the situation in Italy, Paoli invited them to move to Argentina.[48] Corsini's anti-fascist sympathies found some outlet in the scientific activities of the museum; beginning in the late 1930s[49] he employed—albeit informally to circumvent the racial laws—Pietro

---

46 Given the risks of corresponding directly with individuals regarded as *persone non grate* by the regime of Mussolini, Paoli served as an intermediary between Corsini and Mieli.

47 Luigi Castaldi was a medical historian and openly declared anti-Fascist. Piero Franceschini's commentary on these events sheds light on the highly sensitive issue of the stance taken by different intellectuals under the Fascist government. Franceschini was a student of Corsini and later collaborated with him on the rearrangement of the Medici collections in the museum. In June 1962 commemorative ceremonies were held in the museum to honour Corsini's life and achievements, and after listening to the—in his view overly cool and detached—reminiscences of Luigi Belloni, Franceschini took up his pen to write indignantly: "anyone listening to you, if he did not have other elements on which to judge, might have thought that Corsini's life had unfolded in a mythical Limbo. Andrea Corsini was, certainly, a wise and moderate man, but still always a man who was passionately alive; on the morning of the 14th of August 1944 in the Palazzo Vecchio not only did I listen to his vehement expressions of execration for the iniquities of the Nazi occupation [...]. It remains a well-known fact to many of those who have a good memory: during the years 1925–1935 Luigi Castaldi was, in university circles, one of the most determined opponents of Fascism, and it was in those very years that Andrea Corsini openly reconfirmed his personal esteem for Castaldi". Franceschini to Luigi Belloni, 24 June 1962: Fondo IMSS, *Carteggio. Primi anni di vita del Museo.*

48 "I would be most delighted to be able to do something for you, indeed I do not forget your desire, like that of the honourable Dr. Castaldi, [for] things that can be obtained with a little good will and persistence. Moreover various profs are already settled here with their own university chairs and the sympathetic support of everyone, among them many physicians, and the philosopher Mondolfo, with said chair in the Faculty of Letters at Buenos Aires". Umberto Paoli to Corsini, 7 December 1939: Fondo IMSS, *Carteggio Corsini I.* In the same letter Paoli relayed news on the health of Mieli and made inquiries as to the future of the Istituto di Storia della Scienza.

49 Corsini probably deemed it appropriate to wait until after the death of Ginori Conti in 1939 before beginning to work formally with Pagnini. Ginori Conti served as a *Senatore del Regno* under Mussolini, and was a convinced supporter of the ideals of Fascism at least until the regime—in 1939, just months before his death—nationalized his geothermal plant in Larderello.

Pagnini (1875–1955), who was an expert on scientific instruments,[50] to help catalogue the museum's collections.

During these difficult years Corsini continued his efforts to enlarge the museum's collections, first by incorporating the holdings of the Museo della Specola, of which Cipriani and Pagnini had a close knowledge.[51] The room dedicated to the achievements of Alessandro Volta dates to this period (*Fig. 1.5*), with a display surpassed only by the collection conserved at the University of Pavia, where Volta taught for 25 years. The models of mechanical devices from the collection of the Grand Duchy of the House of Lorraine (which ruled Tuscany after the extinction of the Medici line) and the instruments of Felice Fontana (1730–1805), including his registering barometer, were also acquired by the museum during this period.[52] Corsini entered into negotiations to acquire collections of historic interest from other institutions, such as the university of Pavia,[53] and further valuable exhibits arrived in the form of donations and

50 Pietro Pagnini was an openly declared anti-Fascist. Slightly older than Corsini, he had been a classmate of Marconi's and, after graduating from the University of Florence, worked for the Officine Galileo and then, beginning in 1918, for the Società Elettrocarbonium. A militant member of the Partito Popolare, he opposed fascism from the very outset of Mussolini's rise to power. Pagnini collaborated in the preparation of the exhibit and catalogue of the "Florentine instruments" for the 1929 *Esposizione Nazionale*. It was only after the war, however, that he was able to work officially for the museum, being summoned to update and complete the catalogue of its collections and to supervise the restoration of the damaged instruments. On Pagnini's biography, see the short profile BONELLI, "Pietro Pagnini," *Luce e Immagini* 4 (1953), pp. 1–2. His work on the museum's catalogue is documented in Fondo IMSS, *Carte Pagnini*, 1900–1955.

51 In a letter from Ginori Conti to Corsini dated 22 July 1933, we learn that the instruments were finally moved from *La Specola* to the Istituto di Storia della Scienza, despite the request of the president of *La Specola* to Corsini as director of the institute to proceed only on the condition that undue expense could be avoided. Fondo IMSS, *Carteggio Corsini I.*

52 Further confirmation of these acquisitions can be found in the request for the loan of two items—Galileo's inclined plane and Fontana's barometer—dated 21 February 1939 and sent by the engineer Orazio Oldofredi, director general of the *Mostra di Leonardo da Vinci e delle Invenzioni Italiane* (Exhibition on Leonardo da Vinci and Italian Inventions); Fondo IMSS, *Primi Anni di vita del Museo*. It is curious that Oldofredi's contribution is not acknowledged in the introductory pages of the catalogue *Mostra di Leonardo da Vinci. Catalogo*, Milan, 1939.

53 In July 1937 Prof. Rita Brunetti, director of the Istituto di Fisica "A. Volta" at the University of Pavia, sent Corsini a list of 52 scientific instruments that the institute was willing to cede to the museum for the symbolic sum of 10,000 *lire* (ca. 9,000 Euros). Items of particular interest on the list included: an exemplar of Carl Winter's plate frictional electric generator; microscopes constructed by George Adams; a pair of 'binoculars' made by Galileo;

FIGURE 1.5 *The room with Volta's instruments and other electrical apparatus* (PHOTOGRAPHIC LABORATORY AND ARCHIVE, MUSEO GALILEO, FLORENCE).

legacies. In some cases the acquisitions were the result of fortuitous circumstances, as when at the closure of the 1933 *Fiera del libro* (National Book Fair) the museum received for safekeeping the replicas of two antique printing presses, one dating to the eighteenth century and the other used by the typographer Bernardo Cennini. Due to limitations of space, however, these would not be placed on display until after the end of the war.[54] Other donations during the 1930s include an exemplar of Thomas Edison's phonograph from the businessman and philantropist Egisto Fabbri (1933),[55] an 'antique microscope' donated by Luigi Mazzoni (1933), two images originally transmitted by a pantelegraph donated by Pietro Maracchi (1933), instruments and mechanical devices from the scientific instruments manufacturer Officine Galileo (October 1937), apparatus for chemistry experiments from the Laboratorio di Chimica e Agraria of the University of Florence (June 1938), and a splendid collection of medals originally belonging to Giovanni Aperlo (1939).[56]

In 1939 Andrea Corsini participated, as a member of the Organizing Committee, in the preparation of the *Mostra Medicea*, an exhibition dedicated to the Medici dynasty that was held at the Palazzo Medici Riccardi. For this event the museum lent 32 items, including examples of glassware from the collection of the *Accademia del Cimento*.[57] In the same period Corsini refused a

---

a thermometer from the *Accademia del Cimento*; scientific instruments (including an unusual pocket compound microscope) made by the master English artisan Benjamin Martin; and replicas of various machines. Vasco Ronchi acted as the intermediary in this transaction, but it appears that Corsini's interest primarily focused on certain items that he retained essential to complete the museum's collection. Unfortunately, due to lacunae in the documentation it is not clear whether he actually succeeded in acquiring these items. The list is dated 6 July 1937 and was followed on 30 November 1937 by a letter signed by Rita Brunetti, in which the maximum asking price for the lot is specified. Fondo IMSS, 1924–1977. *Carteggio e altri documenti ufficiali.*

54 Letter by Ginori Conti to Fumagalli, 7 April 1933: Fondo IMSS, 1924–1977, *Carteggio e altri documenti ufficiali.*

55 As an article appearing in *La Nazione* on 5 July 1933 observed: "The precious legacy is the object of much admiration, above all by the foreigners who visit this most interesting Museum".

56 The Aperlo collection consists of 366 medallions commemorating the principal figures in the history of medicine and the sciences. During the confusion following the 1966 flood an unknown number of these medallions disappeared. The on-line catalogue of the collection is available at http://exhibits.museogalileo.it/medals/section/GiovanniAperloHisMedalCollection.html (accessed 19 March 2014).

57 Official loan terms (four typewritten pages), 31 March 1939: Fondo IMSS, *Primi anni di vita del Museo.* See also the catalogue of the exhibition *Mostra medicea: Palazzo Medici*, Florence, 1939, pp. 137–137.

request from the organizers of the *Mostra di Leonardo da Vinci e delle Invenzioni Italiane*, whom he suspected of planning to create a Museo Nazionale della Scienza that would be a direct rival to the museum in Florence.

The outbreak of World War II, the devaluation of the Italian currency, the gradual erosion of funding for the university and local institutions, and the death in 1939 of the museum's most steadfast patron, Prince Ginori Conti, made Corsini's efforts to expand and complete the museum's collections even more arduous, although aid did sometimes arrive from unexpected quarters. For example, as one newspaper of the period reported,

> deprived of the necessary funding that would allow a worthy display of the interesting material and the hiring of personnel suitable to provide surveillance over the rooms, the Museum struggles to survive, supported only by the passionate determination of a handful of persons prepared to fight for its survival. Enthusiastic students of local history, many of them artisans who had recently become interested in the collection, have formed a group of *Amici del Museo* [Friends of the Museum] that will contribute, not only with monetary dues, but also with their work destined for the repair of pieces that have deteriorated and whatever other necessity presents itself.[58]

Due to the looming threat of war and the resulting collapse of the financial sector, in June 1940 the museum was forced to close its doors, although its scientific activities continued. Two meetings of the *Società Italiana di Storia delle Scienze Mediche e Naturali* were organized by Corsini and held in the Palazzo Castellani in 1941 and 1942. The second coincided with the 400th centenary of the death of Galileo and for this occasion an important study by Luigi Castaldi on Galileo's microscopes was published,[59] while the public flocked to see relics of the scientist—two fingers and a tooth—belonging to the Rosselli Del Turco family, that were rediscovered by Corsini and Leoncini and placed on display in the museum.[60]

---

58 Giuseppe Sammicheli, "Un Museo dimenticato", 1941 or 1942: Fondo IMSS, *Ritagli* 39/56. The press cutting is not dated and it has not proved possible to retrace the newspaper in which the article originally appeared.

59 Castaldi, "I microscopi costruiti da Galileo Galilei, 1610–1630: con notizie sugli inizi del telescopio", *Rivista di storia delle scienze mediche e naturali* 24 (1942), pp. i–lxvii.

60 These relics, now the property of Candida Bruschi, are temporarily borrowed by the Museo Galileo and on display in Room VII.

On 1 April 1942 Corsini chose Maria Luisa Bonelli (1917–1981) to work alongside the curator Cipriani, who was close to retirement.[61] It was a traumatic time for the city, however, with the unfolding events of the war distracting attention from the centenary celebrations for Galileo, and a particularly anxious time for the museum. In the month of December the most precious items in the collection, among them the relics of the scientist and the delicate glassware from the Cimento collection, were moved to the relative safety of Villa "Il Gioiello", the house of Galileo near the Pian de' Giullari.[62] The premises of the Palazzo Castellani were used to store the furniture of various university professors who had been forced to leave the city and seek refuge elsewhere.[63]

The Post-war Period and the Resurrection of the Museum

The museum emerged from the war somewhat worse for the wear, but fortunately with no irreparable damage. The mines that had exploded in the neighbourhood destroyed almost all of its fixtures, but most of the instruments had been saved.[64] The windows of the palazzo had been blown out and in 1949

61 Bonelli, who was a professor of Spanish literature at the University of Florence, happened to meet Corsini through the friendship between their respective families, and began collaborating with the museum long before she was nominated to the post of curator. I thank Mara Miniati for providing me with this information.

62 "Florence 14 December 1942, [year] XII° [of the Fascist era]. Memorandum on the instruments packed for transport to the Villa Gioiello in Arcetri. The two telescopes of Galileo, the Proportional compass of Galileo, Galileo's Finger, Galileo's Thermoscope, Galileo's 'armed magnet', Galileo's Escapement mechanism, Table with astrolabe belonging to Galileo, drawing by Vincenzo Galilei [Galileo's son], Two globes (celestial and terrestrial) by Viviani, Armillary sphere by Viviani, Hygrometer of Ferdinando II, Glassware from the *Accademia del Cimento*, Thermometers from the *Accademia del Cimento*, Armillary sphere by della Volpaia, Astrolabes, Telescopes, Solar Clocks, Quadrants, Arabian celestial globe, clock with calendar, signs of the zodiac, etc., Galvanometer by Nobili, Electro-magnet by Nobili, Pacinotti's ring. Everything [packed] into no. 15 medium-sized crates, no. 19 small, 3 large *gabbie* and one small." Two typewritten sheets: Fondo IMSS, 1924–1977.

63 Three lists of furniture belonging to Giovanni Bellincioni, the De Filippi family, and Prof. De Robertis: Fondo IMSS, 1924–1977.

64 In the archives is a price quotation for restoration work dated 26 May 1945, from which we learn that the most serious damage involved "a large geographic globe, crushed in three points, [needing] to be completely reconstructed" and the geographic maps on other globes, including one by Coronelli. The total amount specified in the museum's request for funding was 63,300 Lire (ca. 2,200 Euros). Fondo IMSS, *Carteggio. Primi anni di vita del Museo*.

Corsini wrote to the director of the Genio Civile (municipal civil engineering department) with an urgent request for repairs.[65] The exact date is not known, but certainly before April 1946 the great armillary sphere constructed by Antonio Santucci (d. 1613), painstakingly restored, was brought back in triumph to the museum and placed on display on the first floor.[66] In 1946 the museum was re-opened to the public, with the same opening hours as before and the price of an entry ticket fixed at 20 *lire* (equivalent to 60 Eurocents).

With the liberation of Italy in April 1945, Corsini—who was by now 70 years old, but still indomitable—could finally complete the project that he had initiated some twenty years earlier and plan the definitive layout of the exhibits in the Museo di Storia della Scienza. This phase began with the unveiling in 1948 of the museum's splendid globe room, in which pride of place was held by the sphere of Santucci and on the walls of which were hung the fine nineteenth-century stained glass cases that had been rescued from the *Tribuna di Galileo* (*Pl. 1*). Another point of interest was the new library, which contained the institute's valuable collection of books and manuscripts and its collection of busts (*Fig. 1.6*).[67] The first floor of the palazzo was completely restored and divided into eight rooms, although it was decided that repairs to the ground floor should wait.[68] In 1948 the museum hosted the first *Convegno italiano di storia della scienza*, a two-day event during which Pagnini and Bonelli were able to present the museum's refurbished exhibits to the participants.[69] The same premises witnessed the formation, on 7 June 1948, of the *Gruppo Italiano di Storia della Scienza* with the nomination of Corsini by general acclaim as vice-president.

---

65 Letter by Corsini to the director of the Genio Civile, 3 April 1949: Fondo IMSS, *Carteggio. Primi anni di vita del Museo*.

66 After praising Corsini for having "made provision to save and find a fitting place for the sphere of Santucci," Giorgio Abetti proposed that the university transfer to the Palazzo Castellani the six large glass cases of the *Tribuna di Galileo*. Letter by Giorgio Abetti to Piero Calamandrei (the chancellor of the University of Florence), 9 April 1946: Fondo IMSS, *Carteggio. Primi anni di vita del Museo*.

67 A photograph of the globe room and one of the library accompanied a short essay by Bonelli, "Il Palazzo Castellani sede del Museo di Storia della Scienza", *Firenze e il mondo: rivista bimestrale di arti, letteratura e turismo* 1 (1948), pp. 70–72.

68 "All of this material with its valuable exemplars is today arranged in eight rooms on the upper floor and in those rooms on the ground floor that we have to date, albeit only in part, been able to restore." Bonelli, "L'Istituto e Museo di Storia della Scienza di Firenze", *Archives Internationales d'Histoire des Sciences* 6 (1949), pp. 452–456: 455.

69 Sebastiano Timpanaro, "Il convegno italiano di storia della scienza", *Archives Internationales d'Histoire des Sciences* 6 (1949), pp. 460–469.

FIGURE 1.6 *The Library on the first floor of the Palazzo Castellani in 1948* (PHOTOGRAPHIC LABORATORY AND ARCHIVE, MUSEO GALILEO, FLORENCE).

Other developments followed in quick succession. In May 1949 the first *International Conference of Mechanics and Statistics* was held at the museum. Organized by Giovanni Polvani under the auspices of the *Società Italiana di Fisica*, this prestigious meeting saw the participation of distinguished physicists from around the world, including Wolfgang Pauli, Oskar Klein, Max Born and Ilya Prigogine.[70] The museum also began to host concerts in collaboration with Florence's Conservatory of Music, featuring music composed by Vincenzo Galilei (Galileo Galilei's father), and his contemporaries.[71]

In 1950 Corsini managed to acquire the superb Medici-Lorraine library that had been moved to the Istituto di Fisica by Garbasso, a resource that was inseparable from the museum's collection of scientific instruments. Lastly, at Corsini's urging Pietro Pagnini brought the cataloguing of the collections

70 A copy of the printed programme of this international conference is conserved in the Museo Galileo Library.

71 The first concert of which we have any mention was held on 6 April 1949.

almost to completion.[72] Between 1949 and 1951 the museum published six booklets on the history of various instruments: barometers (1949), terrestrial and celestial globes (1950), glassware (1950), telescopes (1950), mechanical instruments (1950), and armillary spheres (1951), each one illustrated with models from the museum's collection.[73] The research conducted by Corsini and Bonelli, as well as Pagnini's publications, attracted the attention of many foreign scholars.[74]

Rival initiatives were not lacking, however. During the *Convegno Nazionale del Gruppo Italiano di Storia della Scienza* which was held at the museum on 27–29 April 1950, Guido Ucelli took Corsini aside in order to reassure him that the soon-to-be-inaugurated Museo Leonardo da Vinci in Milan was not meant to supplant the Museo di Storia della Scienza.[75] All the same, it was Ucelli who revived Corsini's project of compiling an exhaustive inventory of Italy's scientific heritage, which was finally completed in 1961 under the auspices of the Museo Leonardo da Vinci.[76]

---

72 "[...] one day we shall see [them] described in the *catalogue raisonné* [of the Museum] to which Pietro Pagnini has already devoted nearly a year of his valuable work." See Bonelli, "L'Istituto e Museo di Storia della Scienza", cit. (n. 68), p. 456. Interestingly, the work on the catalogue was initially conducted by Pagnini alone. It is difficult to say exactly when Bonelli joined him on the project.

73 The booklets, which were published anonymously but were certainly written by Pagnini, appeared in the first series of scholarly publications by the museum, "Quaderni del Museo di Storia delle Scienze." They comprised: *I barometri*, Empoli, 1949; *Globi terrestri e celesti*, no place indication, 1950; *Vetrerie dell'Accademia del Cimento*, Empoli, 1950; *I cannocchiali esistenti presso il Museo di Storia delle Scienze*, Florence, 1950; *Strumenti di meccanica esistenti presso il Museo di Storia delle Scienze*, Florence, 1950; *Sfere armillari, armille, planetari*, no place indication, 1951. The six works can be downloaded from the virtual exhibition *Museo Galileo 1930–2010*, http://www.museogalileo.it/istituto/en/index.html (accessed 19 March 2014).

74 For example, the monograph issues "Il termometro", *Rivista Ciba* 3, 20 (1949) and "Il microscopio", *Rivista Ciba* 4, 27 (1950) were also published in German and French translations. In 1954 appeared the monograph issue Corsini (ed.), "La medicina alla corte di Pietro Leopoldo", *Rivista Ciba* 8, 46 (1954).

75 Guido Ucelli was an engineer and industrialist (he became director general of Riva Calzoni, producer of large-scale hydraulic machinery) and the founder of the Museo Nazionale della Scienza e della Technologia Leonardo da Vinci. See "Groupes Nationaux: Italie", *Archives internationales d'histoire des sciences* 13 (1950), pp. 898–906: 900.

76 *Inventaire des instruments conservés en Italie*, Milan, 1961, 4 vols.; second enlarged edition: *Inventaire des instruments conservés en Italie*, Milan, 1963, 5 vols.

Corsini bemoaned the disparity between the resources available to the Museo Leonardo da Vinci and the relative penury of his own; museums were perpetually dependent on the patronage of enlightened sponsors and the voluble generosity of a multitude of private and public entities, but the museum in Milan had a much larger pool of donors to draw from. Nevertheless, Corsini continued undeterred in the pursuit of his projects for the improvement of the Museo di Storia della Scienza, which would see in the space of a few years the definitive arrangement of its exhibits.

During the early 1950s the museum holdings grew thanks to private donations and the acquisition of further important collections, beginning with the presentation to the museum of the Mascagni Collection by Federico Allodi and the gift of three Galilean microscopes from Nello Beccari.[77] Corsini also managed to acquire the mineral collection of the Florentine naturalist Giovanni Targioni Tozzetti (1712–1783) (which was, however, eventually returned to the Museo di Storia Naturale of the University of Florence) and the large collection of pharmaceutical jars and surgical instruments from the Ospedale di Santa Maria Nuova.

The expansion of the collections and the preparation of educational exhibits, such as one dedicated to the Cennini printing press, demanded the complete rearrangement of the museum's layout and the preparation of a scholarly catalogue. In the meantime, this first chapter in the history of the Museo di Storia della Scienza came full circle, closing with the publication in 1952 of a catalogue of the famous 1929 exposition. The volume, which came to slightly over one hundred pages and was limited to the scientific instruments in the exhibition,[78] conveyed only a partial impression of what was in truth a historic event, but the cultural matrix of the *Esposizione Nazionale di Storia della Scienza* seemed to have faded and lost its relevance in the post-war age. In all likelihood, the motive for publishing this volume, after a hiatus of nearly 25 years, lay in Corsini's determination to forestall Ucelli and win the credit that was his due for being first to compile an inventory of Italy's scientific heritage. This is suggested by the fact that the catalogue described, in addition to

---

77 The collection includes a microscope constructed in 1811: Fondo IMSS, *Donazioni*. Thanks to donations, between 1950 and 1960 also the museum's archives, library holdings, and collection of prints, photographs and other visual material, were greatly expanded.

78 *Esposizione nazionale di storia delle scienze. Firenze, maggio-ottobre 1929: catalogo con aggiornamenti*, Florence, 1952. The catalogue entries, most of them written by Pagnini, were then edited for publication by Bonelli.

the instruments placed on display in 1929, others that had come to light much more recently.[79]

Corsini's most pressing tasks, however, were to develop a well-conceived didactic tour for visitors to the museum and to finish the work of cataloguing the collections that had been begun by Cipriani and Pagnini. The *Catalogo degli strumenti del Museo di storia della scienza* published in 1954 offered a rich and well-documented overview of the museum's holdings, supported by in-depth archival studies carried out using the latest research criteria. The catalogue proper was authored in the main by Pagnini and Bonelli, but there were scholarly contributions by Federico Allodi and Salvatore Principe on the medical collections, Luigi Belloni on the Aperlo medal collection, Francesco Rodolico on the Targioni Tozzetti mineral collection, and General Federico Negri on the antique firearms collection. The *Catalogo* covered the museum's entire inventory and remains to this day an invaluable reference work and reliable scholarly guide to collections that, for various reasons, have lost the prestigious status that they enjoyed in Corsini's time.[80]

The new arrangement, which took the visitor from the ground floor to the first floor, comprised twenty-four rooms. The first floor, whose layout dated to the late 1940s and was kept more or less unchanged, consisted of eleven rooms, including one for the library and another for the museum's administrative office.[81] On the ground floor, Room I contained various prototypes of the first internal combustion engine, realized by Eugenio Barsanti and Felice Matteucci between 1850 and 1860; Room II pneumatic machines and measuring instruments; Room III various types of weighing scales; Room IV chemical and pharmaceutical instruments; Rooms VI and VII the museum's medical collections; and Room VIII electrostatic machines, while Rooms IX and X were taken up by the two full-scale models of antique printing presses.[82] The magnificent Targioni Tozzetti mineral collection acquired in 1951 was given a worthy setting in the *loggia* on the third floor (*Fig. 1.7*), an open gallery that offered a splendid panorama of the city and was also used to host conferences

79 This is confirmed by Corsini, who wrote: "with […] the catalogue that we have published, i.e., that of the Exposition, we have prepared an initial census of existing historical scientific material in Italy, updated to 1951": *Catalogo degli strumenti del museo di storia della scienza*, Florence, 1954; p. iii.

80 Of particular interest is the exhaustive information that the catalogue provides on the museum's collections of surgical instruments, commemorative medals, chemistry exhibits, and minerals (the mineral collection was returned to the Museo di Storia Naturale in 1990).

81 Bonelli, "Istituto e Museo di Storia della Scienza di Firenze," *Homo Faber* 3, 14 (1952), p. 2.

82 Bonelli, "Il Museo di Storia della Scienza," *Musei e Gallerie d'Italia* 2 (1957), pp. 1–15.

FIGURE 1.7 *The Targioni Tozzetti mineral collection arranged in the* loggiato *on the third floor of the Palazzo Castellani, in a photograph taken in 1951*
(PHOTOGRAPHIC LABORATORY AND ARCHIVE, MUSEO GALILEO, FLORENCE).

and temporary exhibits. One such was an exhibition dedicated to Barsanti and Matteucci's combustion engine.[83] In the same period, thanks to generous funding secured by the distinguished geographer Roberto Almagià (1884–1962) from the Italian National Research Council (CNR), the museum was able to restore some of its most treasured objects, such as Vincenzo Coronelli's globes and Antonio Santucci's sphere.

Sometime before February 1956, having run out of storage space, Corsini set aside two rooms in the basement as a depository for the museum, where some clocks and various more cumbersome items such as models of farming machinery and (possibly) some bicycles and draisines (antique light rail vehicles) were stored (*Fig. 1.8*).[84]

83 Bonelli, "La mostra Barsanti Matteucci," (monograph issue) *La regione* 1 (1954), p. 54.

84 This fact is documented in a letter dated 15 February 1956 and addressed to Niccolò Rodolico, head of the *Deputazione di Storia Patria*, in which Corsini requested the use of space on the third floor, noting that he had recently been forced "even to assign two rooms in the basement to the Museum's use." Fondo IMSS, *Carteggio. Primi anni di vita del Museo*.

FIGURE 1.8 *Antique timepieces and clocks in the museum's collection stored in the basement of the Palazzo Castellani, in a photograph taken in early 1950s*
(PHOTOGRAPHIC LABORATORY AND ARCHIVE, MUSEO GALILEO, FLORENCE).

An extremely fertile period of activity followed that continued to draw impetus from the indomitable enthusiasm of Corsini and his entourage. At the inauguration of the newly-arranged museum and the publication of its catalogue, a glowing article appeared in the *Nazione sera*, which concluded, however, on a warning note:

> It would not be fitting to end this discussion without mentioning the situation of grave want in which the institute finds itself. This museum is unique in Italy because it possesses the originals of historic instruments. In Milan something similar was constituted just a few years ago [in 1953] with the national museum of the history of science and technology, but its collection consists almost exclusively of reproductions of originals that can instead be found on display in Florence, and other models made

> to illustrate various technical points. And yet the collection in Milan, whose value cannot be compared to that of the collection in Florence, has at its disposition the ample premises of the ancient convent of San Vittore, restored by the engineer Guido Ucelli di Nemi after the war and on which the government has disbursed hundreds of millions of lire. The museum in Florence receives each year just six thousand lire from the municipal government, four thousand lire from the provincial government, three hundred and fifty thousand [ca. 5,000 Euros] from the university, and some small contributions from occasional donors.[85]

While it may not be true that the balance sheet of the museum in Florence was quite so dire, the activities that Corsini managed to carry forward with the meagre resources at his disposal are truly remarkable. However, the museum's somewhat involved story does not end here. To guarantee the future of an institution that he hoped would continue to grow, in a strategic move that was also designed to break the deadlock in the negotiations with the *Accademia della Crusca* and the *Deputazione di Storia Patria*, which were still refusing to vacate the Palazzo Castellani, the by-now nearly 80-year-old Corsini commissioned a project to move the entire museum to Galileo's house in Arcetri. This project, which was never implemented,[86] included plans not only for the display of the museum's entire collection of scientific instruments, but also the setting aside of ample space for the institute's other activities, including laboratories, storerooms, a research library, a reading room, and meeting rooms for the hosting of national and international conferences. Presumably the museum's budget would not allow for the realization of such an ambitious project, but other important initiatives could be and were brought to fruition.

With the museum exhibits finally arranged to his satisfaction, Corsini signed an agreement with the publishing house Olschki to collaborate closely on the publication of the *Rivista di storia delle scienze mediche e naturali*, and a series of monographs on topics relating to the history of science. Together with his dedicated assistant Bonelli, Corsini launched a campaign to raise the international profile of the museum, beginning in 1956 with the organization of the *8th International Conference on the History of Science*, which was held

85 Nicola della Santa, *Nazione sera* (30 November 1954); newspaper cutting, Fondo IMSS, *Ritagli*, 39/44.

86 A ground plan with the proposed layout of the museum is conserved in the archives of the Museo Galileo; see Beretta, "Imaging the Museum of History of Science (1841–1991)", in Camerota (ed.), *Displaying Scientific Instruments*…, cit. (n. 25), pp. 151–199: 176–177, figs. 46–49.

at different venues around Italy: Florence, Vinci, and Milan. On this occasion some of the most illustrious science historians in the world visited the museum, where they could admire at first hand the exhibits documenting Italy's unique scientific legacy. Against this international backdrop, in some brief remarks entitled, quite touchingly, *Reminiscenze e rievocazioni di un ottuagenario* (Some Reminiscences and Recollections of an Eighty-Year-Old), Corsini reviewed with pride an exemplary story of endeavour brought to a triumphant conclusion.[87] The weekly magazine *Oggi* published an article illustrated with a generous number of photographs that drew the attention of the general public to this noteworthy institution. A new phase in its history began, with growing popular interest and a rapid rise in the number of visitors.

Upon his death on 14 June 1961, Corsini was universally commemorated for his rectitude and his nobility of soul, for his generosity of spirit and his innate modesty, qualities that have with the passage of the years perhaps tended to obscure and to render more difficult an objective evaluation of his role in the history of the Museo di Storia della Scienza, a role that—as I have sought to illustrate in this brief reconstruction of his lifework—was in reality of exceptional importance from both a cultural and a scientific standpoint.

Thus, a career that began in a small anonymous office in Via degli Alfani came to a happy conclusion. Corsini left a remarkable legacy to the city of Florence and to the international community of scientists and students of the history of science—a thoroughly modern museum and a research institute actively engaged in promoting a greater understanding and appreciation of the scientific and cultural heritage. Through his foresight, determination, and tireless efforts, he created a unique collection of scientific instruments that, with only minor changes, can still be admired today in the recently restored rooms of the Museo della Storia della Scienza, now renamed the Museo Galileo.

---

87 These remarks by Corsini were published in a booklet: *Ai partecipanti all'ottavo congresso internazionale di storia delle scienze*, Rome, 1956, pp. 38–43.

CHAPTER 2

"Not for their Beauty": Instruments and Narratives at the Science Museum, London

Alison Boyle

Introduction

The Science Museum has always worn its name somewhat uneasily. Throughout the institution's history, its staff have expressed anxiety over displaying collections which lack the visual appeal of art objects. The scientific instrument collections in particular have posed challenges for curators in a museum more strongly focused on displays of technology and engineering.

Generations of curators have responded to these challenges by using increasingly overt narrative techniques in the display of objects. Rather than being displayed as stand-alone artefacts to be studied and admired in their own right, the museum's objects have been used to tell visitors stories about science and its history, although periodic attempts to lay out entire collections according to grand narrative schemes have tended to fall by the wayside. Over the years the suite of display tools available to curators has grown, and the collections now jostle for prominence against a variety of exhibition media. Displays of scientific instruments have reflected both internal priorities, for example in the balance of historical and current displays, and external factors, such as the need to nurture industry partnerships and respond to trends in education.

This essay provides an overview of collections displays; more detailed accounts can be found in the Science Museum's official history published in 2010, although the examples in this book focus more on the technology collections.[1] I attempt here to draw out selected examples from the science collections, and to include discussion of more recent decades omitted from this history.

1 Peter J.T. Morris (ed.), *Science for the Nation: Perspectives on the History of the Science Museum*, Basingstoke, 2010. Following the convention used in this book, references to SMD are to the Science Museum Documentation centre, which holds the Museum's institutional archive.

 | DOI 10.1163/9789004264403_003

Scientific Instruments at South Kensington

The name "Science Museum" seems to have first been uttered in 1877, but an independent entity only came into being in 1909, after several decades of efforts to find a suitable repository for the non-art collections of the South Kensington Museum.[2] These collections were envisaged for the use and teaching of artisans, to educate the industrial classes and help Britain catch up with the foreign instrument-making prowess that had become evident from the 1851 Great Exhibition.

In the 1870s the educational collections at South Kensington included "philosophical instruments, apparatus of scientific and other instruction". It was a dynamic collection with items being replaced by manufacturers as they became obsolete, and became a travelling exhibition in the 1870s. It was essentially a trade exhibition for teachers of the sciences, with displays of science textbooks, laboratory fittings and scientific instruments labelled with contact details of the exhibitor and retail prices, and was dispersed in 1889.[3] The lack of a more formal national collection of "instruments used in the Investigation of Mechanical, Chemical or Physical Laws" was lamented in the 1874 report of the Royal Commission into Scientific Instruction. The response was the 1876 Special Loan Exhibition of Scientific Apparatus, overseen by the Commission's secretary Norman Lockyer, astronomer and founding editor of *Nature*. Held in the Museum's Western Arcade and featuring 20,000 scientific instruments from around the world, the exhibition also included displays of historic scientific instruments including Galileo Galilei's telescopes and a quadrant associated with Tycho Brahe (*Fig. 2.1*).[4] At the end of the exhibition, replicas of several historic instruments were made for the Museum collections, perhaps an indication that a complete historical narrative was as important as the education of artisans.

---

2 For a more detailed account of the early history of the South Kensington site, see Robert Bud, "Infected by the Bacillus of Science: The Explosion of South Kensington," in Ibid., pp. 11–40, or David H. Follett, *The Rise of the Science Museum Under Henry Lyons*, London, 1978.

3 *Catalogue of the Educational Division of the South Kensington Museum*, London, 1867; Xerxes Mazda, *The Changing Role of History in the Policy and Collections of the Science Museum, 1857–1973*, MSc thesis, London Centre for the History of Science, Technology and Medicine, 1996.

4 Bud, op. cit. (n. 2), p. 19.

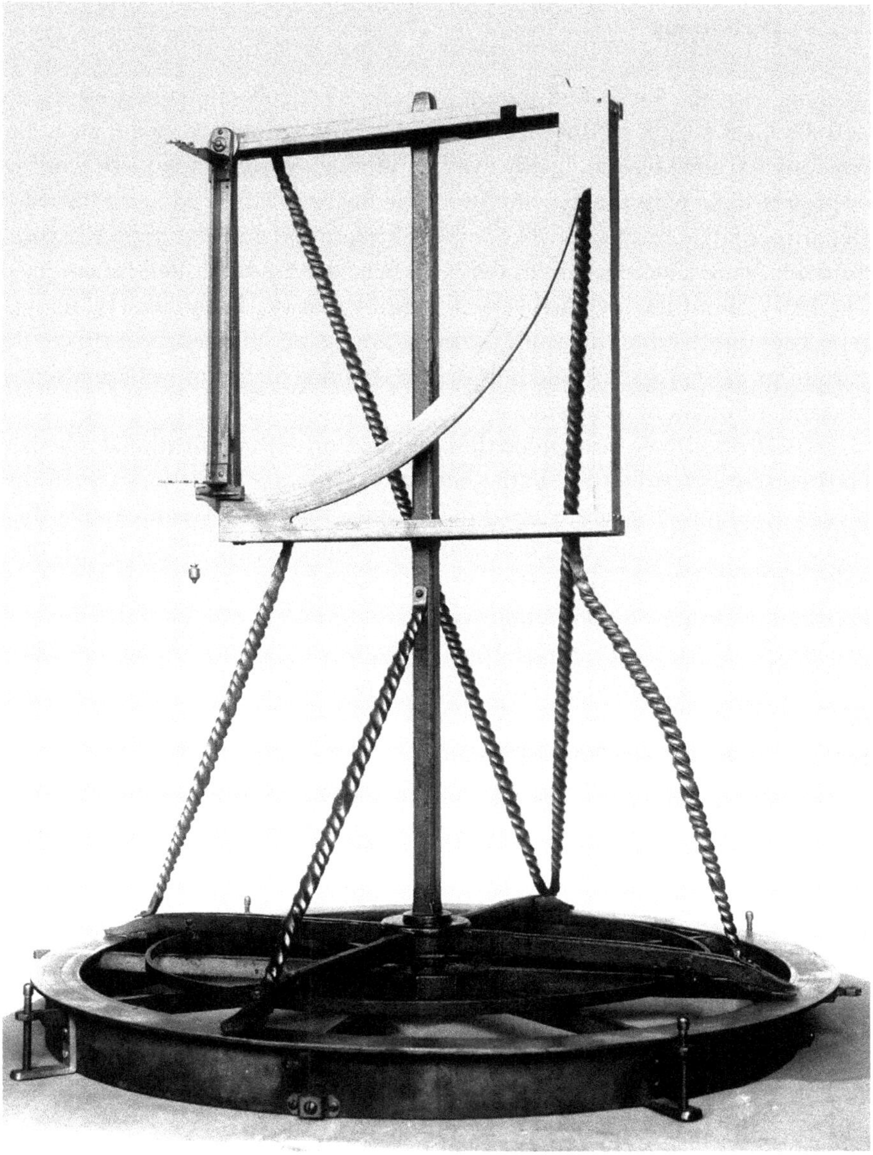

FIGURE 2.1 *Copy of "Tycho Brahe's quadrant" (actually one of the instruments of Landgrave Wilhelm IV of Hessen-Kassel), made for the Science Museum by Elkington & Company before the original was returned to the Staatliche Museen in Kassel, following the 1876 Special Loan Exhibition*
(© SCIENCE MUSEUM / SCIENCE & SOCIETY PICTURE LIBRARY).

Dichotomy

By 1909, with the South Kensington Museum's art collections housed in the new Victoria & Albert Museum buildings across Exhibition Road, an independent Science Museum finally existed, although without adequate funding or premises. To plan the new museum, the Board of Education established a Departmental Committee on the Science Museum and the Geological Museum, normally known after its chair, the steel magnate Sir Hugh Bell (1844–1931). The Bell Committee included several members of the Royal Society. Its 1911 report set out the dual roles of exhibiting the history of science alongside its current practice, which would become the defining dichotomy of the Science Museum throughout its existence.

The Bell Report recommended the continuation of the educational role that had been established by the South Kensington Museum: "so far as is possible by means of exhibited scientific instruments and apparatus, machines and other objects, the Collections in the Science Museum ought to afford illustration and exposition of the various branches of Science within its field and of their applications in the Arts and Industries", and "ought also to be a worthy and suitable house for the preservation of appliances which hold honoured place in the progress of Science or in the history of invention".[5] As Xerxes Mazda has pointed out, the latter recommendation was the first explicit mandate for history of science at the Science Museum.[6]

The Report also recommended that the Museum's displays take the non-specialist visitor into account: "it is of much importance that objects should be so selected and exhibited as to arouse the interest of these visitors, and to afford them in as simple and attractive a form as possible an opportunity of obtaining at least general ideas on the subjects which the Collections illustrate".[7]

In 1911 Francis Grant Ogilvie, a former Director of the Edinburgh Museum of Science and Art who had served as Secretary to the Bell Committee, took over as Director of the Science Museum. At this time the role of the Director was considered small enough that Ogilvie continued to also work as Secretary of the Geological Museum and Geological Survey; his plea to the Board of Education for a raise fell on deaf ears, although they allowed him to appoint an assistant to help run the Museum and its Advisory Council.[8] The setting

---

5 *Report on the Departmental Committee on the Science Museum and the Geological Museum* (also known as the Bell Report), London, 1911, p. 4.

6 Mazda, op. cit. (n. 3), p. 15.

7 Bell Report, cit. (n. 5), p. 9.

8 Follett, op. cit. (n. 2), p. 42.

up of this Council had been recommended by the Bell Committee, several of whose members took seats on the new body. Unsurprisingly, they steered the development of the Museum's collections according to their own recommendations, and secured approval from the Board to begin construction of a new building on Exhibition Road in 1913. However, major development was severely curtailed by the First World War, and even after the War the science displays would suffer: just as the Science Museum began moving its collections into the still-unfinished East Block building, the Western Galleries were given over to the Imperial War Museum and all of the science collections had to be removed to storage.[9] It would take some years for the fledgling Science Museum to establish itself, largely under the influence of Sir Henry Lyons (1864–1944), who took over as Director in 1920 and has been described as "the man who made the Science Museum".[10] Under his Directorship, visitor figures boomed from 400,000 to a million and a quarter annually—more than any other European museum apart from the Louvre. Lyons took the dual ambitions of the Bell Report to heart, and in his plans for the collections we see the start of the Museum's attempts to use narrative schemes to counteract anxiety about the objects not being meaningful in their own right for the general visitor.

Fitness for Purpose

Henry Lyons' first appointment to the Science Museum was in 1912 as Francis Ogilvie's assistant and Secretary of the Advisory Council. By training a military engineer, from boyhood Lyons had also shown a keen amateur interest in geology (pursued so successfully that he was elected to the Geological Society at the age of eighteen). Posted to Egypt in 1890, he combined his army career with geological work and was eventually appointed to direct Geological and Cadastral Surveys of Egypt. Returning to the United Kingdom in 1909, after a short stint as a lecturer at the University of Glasgow, he applied for the Science Museum position. He did not spend long at the Museum before being recalled to military service after the outbreak of the First World War, and became

9 Tom Scheinfeldt, "The First Years: The Science Museum at War and Peace," in Morris, op. cit. (n. 1), pp. 41–60: 41–43.

10 Follett, op. cit. (n. 2), p. 158.

Assistant Director of the Meteorological Office before returning to the Museum in 1919. The next year he was appointed Museum Director.[11]

Lyons' wide-ranging scientific interests, begun from childhood enthusiasm, perhaps underlie his aiming of the Museum's displays even more towards the non-specialist visitor than the Bell Report had recommended. In 1922 he issued a memorandum to his senior staff, defining the Science Museum visitor as first and foremost the 'ordinary visitor', followed in order of decreasing priority by 'the technical visitor', 'the student' and 'the specialist'.[12] He identified the moving of the collections into new accommodation as an opportunity for a comprehensive plan to make it "easier for the visitor to find his way through what must always be a difficult Museum to deal with from the point of view of the visitor as well as from that of the Museum Officer".[13]

The memorandum began by stating that "A Museum of Science and Technology differs essentially in its aim and its central character from an Art Museum [...]. In a technical museum objects are shown on account of their utility, not for their beauty or their attractiveness".[14] The importance of the objects lay in their special fitness for purpose, which would be described in a label or catalogue; the object would provide an illustration of this description rather than being there in its own right. Lyons went on to state that crowding of objects should be reduced, with careful selection so that visitors could look at them with comfort and appreciation.

The science collections were to be displayed in the context of the wider collections: Lyons' aim for the Museum was "to illustrate the influence of science on industry and to show the development which has been achieved thereby", in contrast to other technical museums which might have more specialized or local interpretations of technology.[15] Thus, the pure science collections would be displayed close to collections illustrating industries that had developed from them (for example the new fields of electrical engineering or industrial chemistry) to facilitate cross-reference. Artefacts would be displayed chronologically to "represent the steps by which progress has been made"; periods of arrested or stimulated development should be pointed out. Mazda has noted that the idea of teaching the nature and methods of science through its history

11 Henry H. Dale, "Henry George Lyons. 1864–1944", *Obituary Notices of Fellows of the Royal Society* 4, 13 (November 1, 1944), pp. 795–809.

12 Henry Lyons, "A Memorandum on the Arrangement of the Collections in the Science Museum, to Serve as a Basis for Discussion," October 1922: SMD Z183/1.

13 Ibid., p. 1.

14 Ibid., p. 2.

15 Ibid.

was gaining some currency in educational circles at the time, although it is unclear whether Lyons was aware of this.[16]

Provoking seemingly little enthusiasm from his staff (later noting that "only two or three of the useful ones had joined at that date"), Lyons drove the plan forward largely along these original outlines.[17] In 1923 the collections were regrouped; where formerly there had been one group for machinery and inventions, and one for the science collections, there were now four: industrial machinery and manufacture; mechanical engineering, land transport and construction; water and air transport; and science and scientific instruments under Keeper David Baxandall (1874–1938). For each major collections group the exhibits would feature historical series, based on the permanent collections, and examples illustrating current practice, which would usually be on loan.[18]

Since the East Block building was completed in piecemeal fashion, the initial displays of the science collections did not adhere to Lyons's schemes. A small part of the science collections was redisplayed, drawing on objects from across collections such as astronomy, meteorology, optics, and chemistry, but this was recognised as opportunistic colonising of available space rather than a coherent plan for the science collections.[19] The Museum's Annual Report for 1924 noted that "by the time that these collections can be re-exhibited they will have been for the most part in store for more than eighteen months, and this has very materially reduced the numbers of school parties visiting the Museum, as well as diminished the number of visitors".[20] To speed up the development of Lyons's schemes, scientists were asked to assist in the development of outline schemes: the *Astronomy* section was scoped by the eminent astronomer Harold Spencer Jones (1890–1960), while the distinguished physicist Edward Andrade (1887–1971) was asked to look at a section created by Lyons, called *Properties of Matter and Physical Phenomena*.[21] This caused considerable

16 Mazda, op. cit. (n. 3), p. 17.

17 Follett, op. cit. (n. 2), p. 99.

18 "Science Museum Annual Report for 1923". The reports were published in various formats and this essay adopts the convention for citation described in Morris, op. cit. (n. 1), pp. xx–xxi.

19 "Science Museum Annual Report for 1935".

20 "Science Museum Annual Report for 1924".

21 The Museum's archives only contain initial correspondence with Andrade, and no submitted scopes, so it appears this may have fallen through.

friction between Lyons and Baxandall, who felt that the scientists were contributing little new to the schemes already drawn up by the curatorial staff.[22]

The mutual antipathy of Lyons and Baxandall, whom a colleague later recalled as behaving "like a couple of Kilkenny cats", cannot have helped to further the progress of the science collections displays.[23] A further likely cause for delay is that by the late 1920s Lyons was coming under increasing pressure to align the Science Museum more closely with the interests of British industries. In 1931 the Museum's Advisory Council was reconstituted to increase representation of industrial groups, and from 1933 a series of special exhibitions, largely produced by industry representatives, spotlighted particular contemporary industries.[24] The style of these exhibitions was heavily influenced by the Deutsches Museum in Munich, which had opened to the public in 1925 and was acknowledged as the world-leading science museum; Lyons was part of a Science Museum delegation to Munich in 1929.[25] Ernest E.B. Mackintosh (1880–1957), who took over as Director in 1933, continued the trend. Working exhibits featured prominently in exhibitions such as *Photoelectric Cells and their Applications* (1932) and *Very Low Temperatures* (1936). A notable exception to this style was the pure-research focused *Atom Tracks* (1937), curated by former Cavendish Laboratory physicist Francis A.B. Ward. Here, a single scientific instrument, Charles T.R. Wilson's cloud chamber (1912) loaned by the Cavendish, was surrounded by over eighty cloud track photographs (*Fig. 2.2*). A major aim, ultimately unsuccessful, behind staging this exhibition was the acquisition of the cloud chamber for the permanent collections.[26]

Lyons' grand narratives do not seem to have been realized as far as the science collections were concerned; they seem to have suffered from a lack of available gallery space throughout the 1920s and the Museum's alignment with industry in the 1930s. As hostilities with Germany loomed, planning of new galleries and exhibitions gave way to the practicalities of safeguarding

22 Memos 21–23 January 1925: SMD, ED79/138; Outline scheme for the astronomy section: SMD, ED79/118.

23 Follett, op. cit. (n. 2), p. 102.

24 Scheinfeldt, op. cit. (n. 9), pp. 52–55.

25 Scheinfeldt, "The International Context and the Context of Internationalism", in Morris, op. cit. (n. 1), pp. 294–311: 299.

26 For detailed descriptions of *Very Low Temperatures* and *Atom Tracks*, and the Science Museum's approach to modern physics in this period, see Imogen Clarke, "Negotiating Progress: Promoting 'Modern' Physics in Britain, 1900–1940", PhD Dissertation, University of Manchester, 2012, pp. 64–82.

FIGURE 2.2 *Entrance to the 1937* Atom Tracks *exhibition, featuring C.T.R. Wilson's cloud chamber* (© SCIENCE MUSEUM / SCIENCE & SOCIETY PICTURE LIBRARY).

the collections and much of the material was moved to off-site storage.[27] Mackintosh was conscious that after the war, the Museum should be seen to meet the needs of the beleaguered British public; while staff should not

27 Thad Parsons, "The Science Museum and the Second World War", in Morris, op. cit. (n. 1), pp. 61–89: 62–66.

disregard the importance of maintaining a repository, they must move with the times. Subjects currently popular, such as aeronautics, might merit preferential treatment for a few years: "a compromise seems to be the inevitable answer, with slightly more emphasis on what the public needs (or what we think they need) than on our own individual desires".[28]

New Subjects and a Popular Focus

The Science Museum was fortunate enough not to sustain a direct hit during the war, and on 14 February 1946 reopened part of the East Block to the public. Herman Shaw (1892–1950), the Keeper of Physics and Geophysics who had stood in for Mackintosh while the latter was seconded to war duty, had taken over as Director and realised Mackintosh's plans of giving the public what they wanted: of the twelve galleries which re-opened, eight were given over to an exhibition of wartime aeronautics, which proved very popular.[29] The remaining four galleries were used for a special Science Exhibition that ran until June. This included a section on *Atomic Energy and Uranium*, which seems to have used this newsworthy topic as a hook for displaying items which had been acquired for the young atomic physics collection (e.g. Aston's mass spectrograph, the Cockcroft-Walton apparatus and the cloud chamber photographs that had been amassed in 1937 for *Atom Tracks*). As Parsons notes, the Science Exhibition highlighted a new challenge for the Science Museum following the war: a vast expansion in subjects and material that it could be expected to cover.[30]

The approach of industry-focused temporary exhibitions continued, however. The Science Museum's 1947 celebrations of the 50th anniversary of Joseph J. Thomson's discovery of the electron were "made possible by the whole-hearted co-operation of firms and organisations concerned with the 'electronics' side of electrical industry".[31] The exhibition committee hoped that the visitor "will go away with some appreciation of the way in which a discovery in pure physics, which may at first appear to be of academic interest only, can lead to developments revolutionizing the everyday conditions of civilised

28 Ernest E.B. Mackintosh, "Suggestions for Post-war Policy and Layout", 24 June 1944: SMD Z183/1.

29 Parsons, op. cit. (n. 27), pp. 68–71.

30 Ibid., p. 77.

31 Follett, *The Electron Jubilee Exhibition*, London, 1947, p. 2.

life".[32] The exhibition, which combined objects with working demonstrations, was highly popular, with the entire print run of the exhibition handbook eventually being sold (*Fig. 2.3*).[33] The star object was of course Thomson's cathode-ray tube, but most of the exhibition focused on applications of electron physics research; exhibits illustrated the principles of thermionic valves and cathode ray tubes, noting their applications in radio and other fields. A section on *The Electron as a Wave* included a new acquisition, George P. Thomson's original electron diffraction photographs and camera, alongside a working exhibit of a modern electron diffraction camera for industrial uses.

The 1947 Annual Report noted that aside from the *Electron Jubilee* exhibition, which took up most of the staff time, the year's work "was of the nature of steady progress rather than spectacular achievement".[34] Major redisplays of the permanent collections were some way off, although Shaw had been considering these even before his appointment as Director. In a 1933 memo to Mackintosh, he suggested that the collections form three large groups in the main sections of the building: a Hall of Science, a Hall of Transport and a Hall of Industry and Engineering. The Hall of Science would form the entrance to the Science Museum, where the visitor would "be introduced to the important discoveries of the most outstanding scientists of all ages", which would "be explained and interpreted in an attractive manner readily understood by all".[35] However, Shaw had no time to realize his vision, dying in post in 1950. A lack of long-term leadership of the Museum would be a hallmark of the next few years, hampering plans for permanent displays.

Space for Science

Shaw was succeeded by Dr Frank Sherwood Taylor (1897–1956), Curator of the Museum of the History of Science in Oxford. Unlike his predecessors, he was neither a career civil servant nor a promotion from within the internal ranks. Although Sherwood Taylor was the first trained historian of science to head the Science Museum, he was also a populariser of science. In this view, the institution would "bring before the public, in a pleasing and intelligible form, the principles and practice of science as such and of the particular branches

32 Ibid.

33 "Science Museum Annual Report for 1947".

34 Ibid.

35 Herman Shaw, "A Memorandum on the Future of the Science Museum", 19 June 1943: SMD Z183/1.

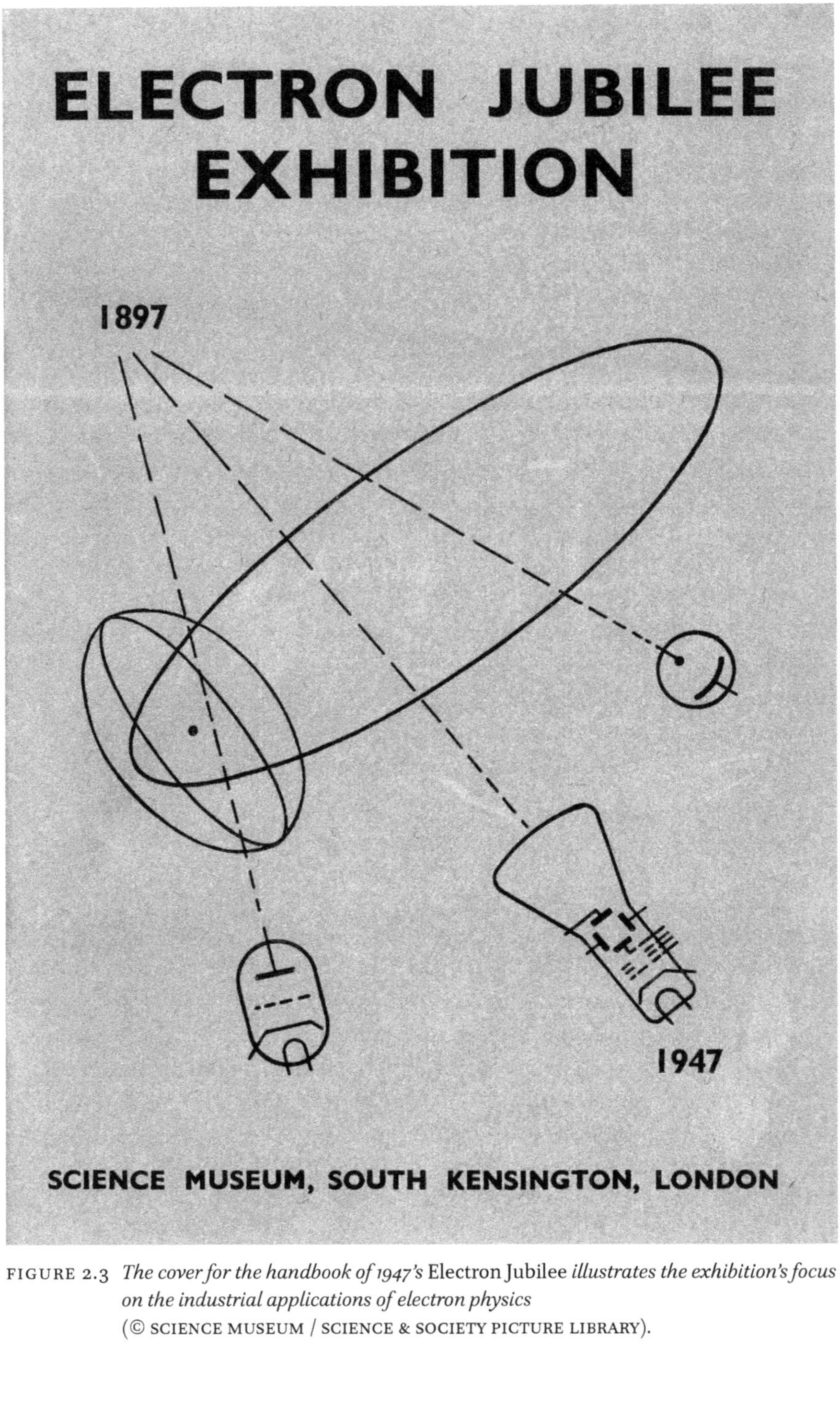

FIGURE 2.3 *The cover for the handbook of 1947's* Electron Jubilee *illustrates the exhibition's focus on the industrial applications of electron physics*
(© SCIENCE MUSEUM / SCIENCE & SOCIETY PICTURE LIBRARY).

of science within the Museum's field, the applications of these sciences to the Arts and Industries, and the total effect upon the life of man".[36]

The 1951 Exhibition of Science for the Festival of Britain, hosted by the Museum, saw an overtly narrative approach. The Festival organisers envisaged this as providing a more detailed exposition of fundamental science than the main Dome of Discovery on the South Bank, but it was designed to complement that exhibition, which borrowed techniques from the commercial and entertainment sectors. Exhibition wallpaper and furnishings featured the crystallography-inspired designs of the Festival Pattern Group; visitors entered under an atom symbol. And, influenced by the Palais de la Decouverte in Paris, films, models and push-button displays were as important as original artefacts. The Festival's exhibits were scripted. At the Science Museum, visitors explored the structure of matter by moving through a series of rooms, exploring at ever-smaller scales. Jacob Bronowski's guidebook adopted the conceit of visitors shrinking like Alice in Wonderland, in Lewis Carroll's famous story. They would imagine themselves the size of a pencil, then down to the layers of lead, to graphite crystals, and finally to the atomic level.[37]

Although the Exhibition of Science was largely curated by the Festival committee rather than the Museum's own Keepers, it had an important legacy for the display of the Museum's collections by giving impetus to the building of the Centre Block. This offered a renewed opportunity to re-organise the galleries by broad subject groups. While the importance of original objects was recognised, they were also noted to have disadvantages in that they were irreplaceable (therefore not allowing sectioning, or other damaging display techniques) and potentially of great size and weight. It was envisaged that the museum would make increased use of working models, dioramas, photographs and full sized reproductions.[38]

Correspondence between Sherwood Taylor and his Keepers about the long-term displays of the collections shows that there was considerable disagreement between the science and technology curators over how the collections should be displayed. Engineering Keeper William T. O'Dea—riding high on the success of the Exhibition of Science and new electrical lighting techniques in the Agriculture gallery—advocated a cohesive display policy, and argued that

36 "Science Museum Annual Report for 1952".

37 Sophie Forgan, "Festivals of Science and the Two Cultures: Science, Design and Display in the Festival of Britain, 1951", *British Journal for the History of Science* 31, 2 (1998), pp. 217–240.

38 "Science Museum Annual Report for 1952".

science objects, being 'so very small', only merited 25% of the display space.[39] Ward and Henry R. Calvert, the Keeper of Astronomy and Geophysics, argued hotly against the space restrictions, pointing out the expansion of the sciences into new areas such as atomic physics and industrial chemistry, and noting the existence of large objects such as telescopes and turret clocks. Calvert was not overly impressed with O'Dea's new display style: "We seem at the moment to be going all-out for dioramas. I think this can be overdone. They are only one stage better than an oil painting and two stages better than a photographic print. What we ought to show is the original historic material against an attractive and appropriate back-ground".[40] Calvert also fought a battle with O'Dea over the prime top-floor gallery in the new Centre Block; which ultimately was given to Aeronautics, while Astronomy was demoted to the first floor, away from the rooftop observatory.[41]

While the arguments over space and display were going on, the curators were struggling to meet Sherwood Taylor's brief to create a stand-alone section on Science and Scientific Method. Ward, who admitted in a later memo to Sherwood Taylor that he considered the Museum an unsuitable forum to cover the scientific method, outlined a scheme illustrating 'scientific method' as a number of distinct approaches rather than a single invariable approach.[42] He identified seven approaches:

A. Simply noticing things hitherto unobserved, or, if observed, not interpreted.
B. Sheer investigation and measurement of phenomena
C. Orderly collection and marshalling of existing facts, leading to new generalisations and further applications of these generalisations
D. Successful application of the phenomena typical in one branch of science to another branch where similar considerations apply
E. Application of new techniques to old problems
F. Investigation of causes of small discrepancies revealed by accurate measurement.

39 For more detail on O'Dea's display approaches see David Rooney, "A Worthy and Suitable House: The Science Museum Buildings and the Temporality of Space", in Morris, op. cit. (n. 1), pp. 157–175: 158–170, and Andrew Nahum, "Exhibiting Science: Changing Conceptions of Science Museum Display", in Ibid., pp. 176–193: 180–186.

40 Henry R. Calvert, "Space Allocations", 2 October 1951: SMD Z183/1.

41 Rooney, op. cit. (n. 39), p. 169.

42 Francis A.B. Ward, "Memo to Director Regarding Long-term Plans", 23 December 1952: SMD Z183/1; Ward, "Scientific Methods in General. Possible Representation in the Museum", 1 December 1952: SMD ED79/143.

Illustrating his notes with "a somewhat random list of important discoveries or steps in branches of science" coming within the scope of the physics department, Ward proposed that Becquerel's discovery of radioactivity was *A*, Millikan's measurement of the charge of the electron was *B*, and Chadwick's discovery of the neutron was *C*. G.P. Thomson's proof of the wave nature of the electron—analogous to the demonstration of the wave-like nature of X-rays by diffraction rings—was *D*, Cecil Powell's application of photographic techniques to cosmic-ray research was *E*, while Curie's discovery of radium was *F*. Historical examples did not always fit neatly into distinct categories: J.J. Thomson's discovery of the electron was *C* arising from *B*, while Carl Anderson's discovery of the positron using a Wilson cloud chamber was *A* arising from *E*. Theoretical physics seems to have come under *C*: Newton's laws of motion and Kelvin's laws of thermodynamics were both identified as this, while Rutherford's nuclear atom was identified as *C* arising from *B*. Ward was of the opinion that *A*, *B* and *E* were already well-represented by the Museum's exhibits, while *C* and *F* did not lend themselves to demonstration by three-dimensional museum objects.

Another scheme, drawn up by G.F. Westcott, one of the Engineering keepers, got around the problem of defining scientific method by re-defining the collection as "Science, its (History), Methods and Relations with Society", presented via a series of dioramas illustrating historical scenes, but deliberately eschewing any explicit definition of scientific method.[43] Sherwood Taylor eventually abandoned the plans due to lack of space, leaving scientific method "to be thought of by some future director and body of keepers".[44]

The Rise of History

In 1960 the Science Museum reverted to its policy of promoting an internal candidate: David Follett (1907–1982), who by then had moved from the physics department to be Keeper of Electrical Engineering and Communication. As Mazda has noted, Follett re-aligned the Museum to expressly promote the history of science and industry, rather than using history as a tool to explain science.[45] Follett's paper to the Advisory Council in June 1960 saw

43 G.F. Westcott, "Proposed Scheme for a Collection Illustrating Methods of Science", 1953: SMD ED79/143.

44 Frank Sherwood Taylor, "Memo on Schemes for Development, Science & Scientific Methods 1952–53", 5 August 1953: SMD ED79/143.

45 Mazda, op. cit. (n. 3), p. 24.

the occupation of the new Centre Block as an opportunity to re-organise the collections, allowing the Museum "take its place on equal terms among the other famous museums and galleries as part of the cultural background of the nation". He set the specific objectives of developing the Museum as a centre of historical research (publishing in journals, rather than just writing historical surveys for collections handbooks) and developing displays to reflect this, noting that the existing practice of displaying objects alongside dioramas and contemporary prints was a rather sporadic way of providing social historical background.[46]

Like Lyons before him, Follett was unconvinced of the visual appeal of the Museum's collections: "Unlike the exhibits at, for example, the National Gallery, or the Victoria and Albert Museum—paintings, sculpture, aesthetics and so on, most scientific objects have no aesthetic appeal; they are of interest only in the context of their origin and role".[47] He advocated "a change from the present system whereby each officer does the best he can, according to his inclinations" in favour of working with professional display artists.[48]

Follett's historical and design approach was well illustrated by the Chemistry galleries (1964 and 1977), which combined chronological instrument displays with large-scale recreations of laboratories including Agricola's alchemical workshop (disingenuously described as an 'assaying laboratory') and the 1895 Government Chemist's Laboratory (*Pl.* 2). Arguably the highlight for many visitors was an enormous interactive periodic table, designed by an external firm.[49]

Narrative as King

In the later decades of the twentieth century, with the availability of a wider variety of display techniques and the increasing influence of professional educators, the instruments themselves were frequently displaced as the primary carrier of the exhibition narrative. Andrew Nahum notes that there was sometimes tension between storytelling and collections: the technique of planning

46 Follett, "The Presentation of the Museum's Collections", 31 January 1961: SMD Z183/2.

47 Follett, "Briefing for Minister's Speech at the Opening of the New Aeronautics Gallery," 25 June 1963: SMD Z183/2.

48 Follett, "The Presentation of the Museum's Collections", cit. (n. 46).

49 Morris, "The Image of Chemistry Presented by the Science Museum, London in the Twentieth Century: An International Perspective," *Hyle* 12, 2 (2006), pp. 215–234.

an exhibition by its 'messages' did not always map onto existing collections, which were acquired in a different age and possibly with different concerns. The 'strong' versions of this technique saw ruthless pruning of exhibits to the minimum number that could deliver the messages.[50]

Follett's drive for professionalism in Museum displays saw the setting up of Museum Services in 1967, with Margaret Weston as Keeper. Weston had trained as an electrical engineer and joined the Museum in 1955 as an Assistant Keeper working on the new Electrical Engineering gallery.[51] Speaking with Scott Anthony in 2009, she recalled that Follett had recognised her capabilities and offered her opportunities working on the Children's Gallery followed by Museum Services; while these may have been seen as less prestigious than the subject-specialist curatorial positions, she believed that the opportunity to gain broad experience of Museum operations served her well when she rose to the position of Director.[52]

Many staff recruited to Museum Services took a self-conscious interest in museology, and incorporated techniques from the new American science centres.[53] Weston became Director in 1977, and while the most influential gallery of her tenure was the science-centre style Launch Pad gallery (opened to the public in 1986 after her retirement from the Museum), the increasing dominance of thematic storytelling and new display techniques was also seen in the object-rich galleries. *Exploration* (1977) saw instruments from diverse collections such as space technology and biomedicine displayed under a broad unifying theme, while in galleries such as *Nuclear Physics* (1982) graphic panels ordered the visit, rather than being optional add-ons to the showcase displays.

The *Chemical Industry* gallery, opened in 1986, was the strongest example of this approach, with the objects housed in a design echoing a chemical works. Nahum has noted this gallery as "the emergence of the desire [...] to create exhibitions which are designed as total architectural spaces in support of the narrative or 'spirit' of the exhibition".[54] Gallery curator Robert Bud has described how, in the negotiations between the museum and external design team, important but inconveniently large objects had to be cut, requiring

---

50 Nahum, op. cit. (n. 39), p. 189.

51 "Weston, Dame Margaret (Kate)", *Who's Who 2014*, Online Ed., http://www.ukwhoswho.com/view/article/oupww/whoswho/U39427 (accessed December 2013).

52 Scott Anthony, "Ambition and Anxiety: The Science Museum, 1950–1983", in Morris, op. cit. (n. 1), pp. 90–110: 103–104.

53 Ibid., pp. 102–103.

54 Nahum, op. cit. (n. 39), p. 188.

regular revisions of the gallery story to ensure that the overall narrative logic was maintained.[55]

Two Museums

During the late 1980s and 1990s the rise of the United Kingdom's Public Understanding of Science (later Public Engagement with Science) movement saw the Museum focus heavily on 'issue led' exhibitions, often to the exclusion of instrument displays. In 1989 John Durant joined the Museum as Assistant Director and Head of Science Communication. With a background in biological sciences and history of science, Durant simultaneously held a Professorship of Public Understanding of Science at the Museum's neighbour, Imperial College.[56] In 1992 he identified two aims for science museums: the holding of collections, and the promotion of public understanding of science via exhibitions and associated programmes (an aim also fulfilled by collection-free science centres). He argued that the collections focused on the past and the practical, rather than the current or theoretical, admitting that "as often as not the artefacts of twentieth-century science are both mind-bogglingly dull and utterly uninformative to behold".[57] He welcomed a blurring of distinctions between science museums and science centres, with genuine scientific apparatus couched within highly interactive exhibitions. However, for most of the 1990s and 2000s the reality as played out on the gallery floors was more akin to two different institutions—a science museum and a science centre—sharing space within the one building.

Under the Directorship of Neil Cossons (born 1939 and Director from 1986–2000), the Museum was reorganized, creating a split between the Collections Management Division, including the curators, and a Public Services Division, responsible for interpretation, presentation and education working alongside the curators. Later, a separate Science Communication Division was created. Timothy Boon has noted that this created a mixed economy of gallery

55 Bud, "The myth and the machine: seeing science through museum eyes", in Gordon Fyfe, John Law (eds.), *Picturing Power, Visual Depiction and Social Relations*, London, 1988, pp. 134–159: 146–152.

56 "Durant, Prof. John Robert," *Who's Who 2014*, Online Ed., http://www.ukwhoswho.com/view/article/oupww/whoswho/U14394 (accessed December 2013).

57 John Durant (ed.), *Museums and the Public Understanding of Science*, London, 1992, p. 9.

production at the Museum.[58] For example *Food for Thought* (1988), prepared by members of the Public Services Division, contained a series of room sets in which objects were essentially props, and controversially displaced the Astronomy collection to storage, where it would remain for the next two decades.[59] Meanwhile, the curator-led *Science in the 18th Century* (1993) put the objects to the forefront. Arguably the King George III collection, with its wonderful demonstration apparatus, was a rare instance where the Science Museum could display objects 'for their beauty', although the gallery curators had originally hoped for more illustrations, demonstrations and possibly room sets than could be accommodated in the rather sparse final design.[60]

The separation between the curators and science communicators was physically realised in the Science Museum's architecture with the 2000 opening of the Wellcome Wing, devoted to contemporary science, side-by-side with the *Making the Modern World* gallery, showcasing the cream of the Museum's collection (*Figs 2.4 and 2.5*). The Wellcome Wing, with gallery interpretation developed according to learning theory, relied heavily on graphics and new media displays. Galleries were planned according to a hierarchy of messages, which individual exhibits would contribute to. This was at its most extreme in the magazine-style *Antenna* science news gallery, where the story led and objects were frequently used merely as illustrations, if they appeared at all. *Making the Modern World* (MMW) was a chronological display of science and technology artefacts from 1750–2000 and led with the objects, using a 'gestalt' approach with minimal interpretation. The curators envisaged that on entering the gallery, with its stunning object panoramas, visitors would get a holistic sense of the development of science and technology over the past 250 years and of the huge range of the Museum's collections. The gallery's main narrative strands—icons of invention, thematic displays (e.g. measurement, mass production, growing public ambivalence) and technology in everyday life—were laid out according to a chronology flowing the length of the gallery. Within these strands, interpretation of individual objects was minimal: a label, occasionally accompanied by a small illustration or film.[61]

58 Timothy Boon, "Parallax Error? A Participant's Account of the Science Museum, c.1980–2000", in Morris, op. cit. (n. 1), pp. 111–135: 125–126.

59 Allan Chapman, "How Long Must Heaven Wait? At the Science Museum, Astronomy Is Out and Fast-food In", *The Independent*, 25 October 1992.

60 Jane Wess, Personal Communication, 14 October 2013.

61 Ben Russell, "Representing the History and Contemporary Practice of Technology at the Science Museum, London", MA Thesis, University of Leicester, 2006, pp. 34–41.

FIGURE 2.4 *The Wellcome Wing, opened in 2000*
(© SCIENCE MUSEUM / SCIENCE & SOCIETY PICTURE LIBRARY).

These very different approaches met with mixed success. Evaluation, admittedly on the model of learning used in the development of the Wellcome Wing galleries, showed that *MMW* presupposed too high a level of knowledge on the part of the average visitor for the 'gestalt' approach to work. Most engaged with 'iconic' objects such as Apollo 10, and enjoyed the nostalgia of displays of recent technology, but struggled to place unfamiliar objects in context,

FIGURE 2.5 Making the Modern World, *opened 2000. J.J. Thomson's cathode ray tube is amongst a group of instruments in the showcase at the front of the image* (© SCIENCE MUSEUM / SCIENCE & SOCIETY PICTURE LIBRARY).

particularly for the earlier chronological displays.[62] It could also be argued that the scientific instruments on display in the gallery were overshadowed by the large technology exhibits, particularly as the gallery architecture tended to

62 Ibid., pp. 41–43.

funnel visitors along the central aisle. Meanwhile, the Wellcome Wing's *Who am I?* gallery, with its focus on contemporary genetics and brain science, was found to be a highly effective learning environment, but not wholly successful in engaging visitors with artefacts: the interactive displays and object showcases were separately situated and the interactives tended to dominate visitor attention.[63]

Making the Modern World was the final object-rich gallery to be opened for some years. Under the Directorship of Lindsay Sharp, the contemporary science approach dominated, with openings including *Antenna* displays, the *Energy: fuelling the future* interactive gallery, and the Dana Centre for the discussion of current science issues. Audience research indicated that these enterprises were popular with visitors and generally successful in delivering their educational aims.[64] However the Museum's role in the care and interpretation of its historic collections was somewhat sidelined during this period: the museum group manifesto, while stating that more collections should be on display, made little explicit mention of the collection.[65]

Best of Both Worlds

In recent years the Science Museum has sought to blend the best of the two polar approaches, developing object-led galleries with a strong narrative, using a range of display techniques to support the objects rather than compete with them. The drive towards narrative, and the ever-increasing blurring of scientific disciplines, has seen a move away from permanent galleries based on narrow academic disciplines or single collections. The period between 1997–2007 saw the closure of galleries such as Weights and Measures, Nuclear Physics and Power, Chemical Industry, Optics, Photography & Cinematography, Heat & Temperature, and Oceanography.

The loss of display space for the science collections during this period has not yet been rectified; Robert Bud has noted that there were fewer scientific artefacts on display at the Science Museum at the beginning of the twenty-first century than had been displayed in the 1876 Loan Exhibition.[66] Long-term ambitions for a *Making of Modern Science* gallery above *Making the Modern World* tied up the science curators for years; the many different display schemes

63 Ibid., p. 29, table 3.2.

64 Internal Science Museum reports are available on request.

65 *In the 21st Century What Role Should a Museum Play?*, London, 2002.

66 Bud, op. cit. (n. 2) p. 19.

tabled attest to the difficulties of developing a synoptic gallery on the history of science, just as the curators had found with *Science and Scientific Methods* in the 1950s—indeed, the scheme finally chosen to be designed and built had more than a nod to Westcott's despite the gallery team being unaware of his work.[67] The gallery was finally shelved by Director Ian Blatchford in 2010, on the grounds that the history of science could not be satisfactorily displayed in a single gallery, especially when this should be the concern of the whole Museum. At the time of writing, plans are underway for a new series of permanent science galleries but delivery is some years off.

In the interim, temporary exhibitions of the science collections have offered an opportunity to try out new techniques. *Cosmos & Culture* (intended as a temporary gallery in 2009, but still in situ) eschewed panels and labels, instead using multimedia screens to interpret the astronomy collections. While these offered the possibility to include more historical context and explanations of instrumental workings than could be provided by printed interpretation, the lack of an overall graphic structure and the grouping of objects inside very large showcases meant that visitors could not discern the gallery themes very well.

Artefacts from the Large Hadron Collider (LHC) arguably embody Durant's critique of twentieth century science, and recent exhibitions on particle physics have failed to engage audiences with the objects, or provide the necessary sense of scale.[68] To address these problems, the Science Museum is experimenting with theatrical collaborations in a temporary exhibition, *Collider* (November 2013–May 2014). LHC artefacts are displayed in sets resembling different spaces at CERN in an attempt to set them in the context of where they were made and used. The narrative follows the journey of the LHC's particle beam, with each object displayed at the beam stage it was designed for. J.J. Thomson's cathode ray tube has a dual identity in the exhibition: it sits in the introductory area as part of a 'classic' object display exploring themes in the history of particle physics, while an opening in the back of its showcase allows it to also form part of a screened drama with a scene of Thomson and his assistant Ebenezer Everett at the Cavendish laboratory in 1897: here it is more of a prop, albeit a prominent one (*Fig. 2.6*). Thus it both guides a narrative, and has a narrative imposed on it; the ambivalent role of scientific instruments in the Science Museum continues.

---

67 SMD, SCM/2011/0943.

68 Marie Hobson, *LHC and CERN: A Summary of the Public's Knowledge, Understanding and Interest*, Science Museum Internal Report, February 2012.

FIGURE 2.6 *Two display modes of J.J. Thomson's cathode ray tube (1897) in the* Collider *exhibition, 2013* (© SCIENCE MUSEUM / SCIENCE & SOCIETY PICTURE LIBRARY).

CHAPTER 3

"More Artistic than Scientific": Exhibiting Instruments as Decorative Arts in the Victoria and Albert Museum

Richard Dunn

Introduction

It is perfectly common to find scientific instruments, broadly construed, in museums of the decorative or applied arts and even to find them displayed there.[1] This is the context examined in this essay, through an exploration of the history of the Victoria and Albert Museum, London (V&A).[2] In doing so, it also considers some of the more obvious differences between displays of the decorative arts and of science, evident in the contrast between the V&A and the Science Museum, two very different museums with a common origin.

The South Kensington Project

As has been well discussed, the V&A and the Science Museum began life as a single institution, the South Kensington Museum, which was conceived as the lasting legacy of the Great Exhibition of 1851 and opened to the public in

1 For example, Mary Holbrook et al., *Science Preserved*, London, 1992, lists many examples of instruments in different types of collection in the United Kingdom. On the nature of instruments in museums, see Judith V. Field, "What is Scientific About a Scientific Instrument?", *Nuncius* 3 (1988), pp. 3–26; Deborah J. Warner, "What is a Scientific Instrument, When did it Become One, and Why?", *British Journal for the History of Science* 23 (1990), pp. 83–93, and Liba Taub, "On Scientific Instruments", *Studies in History and Philosophy of Science* 40 (2009), pp. 337–343. On the ambiguity of terms such as 'decorative arts' and 'applied arts', see Charles Saumarez-Smith, "The Classification of Things", *Bulletin of the John Rylands Library* 77, 1 (1995), pp. 13–20.

2 For the instruments at the V&A, see Richard Dunn, "Scientific Instruments at the Victoria and Albert Museum, London: A Provisional Inventory", *Bulletin of the Scientific Instrument Society* 79 (2003), pp. 6–14, and "Corrections", *Bulletin of the Scientific Instrument Society* 84 (2005), p. 1.

 | DOI 10.1163/9789004264403_004

1857.[3] As the showcase of the Department of Science and Art, the new museum's collections and displays exemplified, at least in principle, the idea that science and art could and should be considered together for a single goal: the improvement of British manufacturing through the education of designers and artisans.[4] In practice, however, the execution was not as unified as this might sound.

The departure from a coherent ideal reflected the piecemeal genesis of the new museum, which was not formed from scratch but brought together an assortment of mostly pre-existing collections: Ornamental Art (soon re-named the Art Museum); British Art; Modern Sculpture; an Architectural Museum; a Circulating Art Library; Models etc. Illustrative of Building; an Educational Museum (*Fig. 3.1*); Animal Products (later joined by the Buckland Fish collection); and the Patent Museum (which remained autonomous until 1883).[5] Pulling together such a broad mixture of collections was far from straightforward, however. The Museum's founder, Henry Cole (1808–1882), complained that it was little more than a "refuge for destitute collections",[6] although after visiting a writer in the *Leisure Hour* was more generous in his judgement that, "Multifarious the contributions are—an *omnium gatherum*, reminding one of the cabinet of the virtuoso" (*Fig. 3.2*).[7]

Nor did physical adjacency lead to a free flow of information and artefacts between the different sections. Rather, each collection retained a separate identity and acquisitions were made with little attention to what was happening elsewhere in the Museum, although occasional transfers and loans did take place. Indeed, a disciplinary rift soon opened up and by 1860 was publicly acknowledged, when the Museum's guide book drew a distinction between the Art Division (Art Library, Museum of Ornamental Art, Sculpture Collection, Architectural Collection, and Gallery of British Fine Art) and Science Division (Educational Collections, Animal Products Collection, Food

3 Malcolm Baker, Brenda Richardson (eds.), *A Grand Design: The Art of the Victoria and Albert Museum*, New York, 1997; Robert Bud, "Infected by the Bacillus of Science: The Explosion of South Kensington", in Peter J. Morris (ed.), *Science for the Nation*, Basingstoke-London, 2010, pp. 11–40; Anthony Burton, *Vision & Accident. The Story of the Victoria & Albert Museum*, London, 1999; David Follett, *The Rise of the Science Museum under Henry Lyons*, London, 1978.

4 Rafael Cardoso Denis, "Teaching by Example: Education and the Formation of South Kensington's Museums", in Baker, Richardson (eds.), op. cit. (n. 3), pp. 107–116.

5 *Fifth Report of the Science and Art Department*, London, 1858, pp. 55–79.

6 Henry Cole, *Introductory Addresses on the Science and Art Department of the South Kensington Museum. No.1. The Functions of the Science and Art Department*, London, 1857, p. 20.

7 "The South Kensington Museum", *The Leisure Hour* 8 (1859), pp. 215–218: 216.

FIGURE 3.1 *The Educational Collection at the South Kensington Museum in 1859, with the Art Museum visible through the doorway at the back* (© VICTORIA AND ALBERT MUSEUM, LONDON).

Museum and Structural Museum).[8] Given this partition, it is not surprising that different collections within the two divisions were collecting sometimes similar material, including scientific instruments, in the decades before the Science Museum gained independence from the V&A.

Scientific Instruments in the Art Collections

From the start, then, there were distinct contexts for the collection and display of instruments at South Kensington. One of these was the Art Museum, which collected them at an extremely modest but steady rate during the nineteenth century, with the holdings standing at about 125 by 1900. By then, sundials (22), scales and weights (10), astrolabes (9), measuring devices (8) and

8 *Guide to the South Kensington Museum*, London, 1860, pp. 1–6.

FIGURE 3.2 *The Southern Gallery at the South Kensington Museum in the 19th century* (PRIVATE COLLECTION).

barometers (8) formed the largest groups. An emphasis on the 'industrial arts', which the Art Museum was central in promoting, clearly favoured the acquisition of these objects with a practical everyday use. The place of sundials among the decorative arts had also been well-established by the end of the eighteenth century when a series of studies on ancient sundials had been published as a source book of designs for the applied arts.[9] This matched very well the South Kensington Museum's overall educational aims, which have remained largely, though not exclusively, foremost among the art collections.[10]

One of the principal motives for collecting and displaying scientific instruments in the Art Museum, which opened with predominantly materials-based displays, was as examples of fine metalwork or woodwork. Indeed, this was evident even before the opening of the South Kensington Museum. From 1852, a temporary legacy of the Great Exhibition was the Museum of Ornamental Art in Marlborough House, one of the founding collections at South Kensington. Here the displays included items bought at the sales of two antiquarian

9 Anthony J. Turner, "From Mathematical Practice to the History of Science", *Journal of the History of Collections* 7 (1995), pp. 135–150: 143.

10 Richard Dunn, "The Future of Collecting: Lessons from the Past", in Simon J. Knell (ed.), *Museums and the Future of Collecting*, Aldershot, 2004, pp. 62–71.

FIGURE 3.3 *Artillery instrument, German, dated 1617, though possibly of later date* (V&A 2076-1865; © VICTORIA AND ALBERT MUSEUM, LONDON).

collections, those of Ralph Bernal (1783–1854) and Jules Soulages (1803–1857). Each included finely crafted instruments, with an artillery dial (*Fig. 3.3*) and two sundials entering the collections among the purchases. It is noteworthy that the artillery instrument appeared at the Bernal sale among examples of 'medieval metalwork', rather than with the calendars, compasses and dials from which the British Museum acquired several pieces. Moreover, while the sale catalogue described the level as a "small brass almanac and compass", the

Art Museum register later called it an "astronomical instrument or sundial".[11] Ongoing uncertainty about the nature and use of an instrument purchased primarily for its decorative features was evident in the changing descriptions. As this implies, questions of style and decorative technique largely overshadowed any concern for technical knowledge of the instruments and their function, something that was also evident in display labels.

Ironically, it is more difficult to discern straightforward motives in the Science Division in its first years, in particular concerning historic instruments. Xerxes Mazda, for instance, has noted a distinct tension between the pedagogical emphasis of the Educational Collections and any desire to collect and display the history of science (mirroring similar tensions in the Art Division), the latter only gaining prominence from the 1870s.[12] Others clearly perceived this, for as early as 1866, the horological collector, Charles Octavius Swinnerton Morgan (1803–1888), was visibly unimpressed by the Museum's seemingly poor understanding of historic artefacts. In a letter to the British Museum about a mechanical ship-clock, he declared that, "I should not like to see it transferred to the South Kensington Museum, for, from the arrangement of objects there, it seems to me that such matters are not understood and therefore not appreciated".[13]

In the first decades at South Kensington, then, one could argue that historic instruments fitted more easily into the Art Museum's collections and displays, where questions of style and technique gave them a legitimate context, than in those of the Science Division. This situation was changing by the 1880s, however, as the staff of the Science Division began to assert a more explicit interest in historic material in the wake of the 1876 Special Loan Exhibition.[14]

---

11 Henry G. Bohn, *A Guide to the Knowledge of Pottery, Porcelain and Other Objects of Vertu, Comprising an Illustrated Catalogue of the Bernal Collection of Works of Art*, London, 1872, pp. 126 and 403–404; *Inventory of the Objects in the Art Division of the Museum at South Kensington*, Volume I, London, 1868, museum no. 2076–1865.

12 Xerxes Mazda, *The Changing Role of History in the Policy and Collections of the Science Museum, 1857–1973*, "Science Museum Papers in the History of Technology 3", London, 1996; see also Robert Anderson, "Connoisseurship, Pedagogy or Antiquarianism?", *Journal of the History of Collections* 7 (1995), pp. 211–225: 218–220, and Bud, op. cit. (n. 3), pp. 17–23.

13 Quoted in Anderson, op. cit. (n. 12), p. 223.

14 Bud, op. cit. (n. 3), pp. 17–23; Bud, "Collecting for the Science Museum: Constructing the Collections, the Culture and the Institution", in Morris, op. cit. (n. 3), pp. 250–272: 254–259.

Separating Art and Science

By the end of the nineteenth century, the two halves of the South Kensington Museum were growing further apart. Separate directors were appointed in 1893, and in 1909 the Science Museum's name was formally separated from that of the Victoria and Albert Museum (its official name since 1899), with the two sites physically separated by Exhibition Road. As the two institutions grew into their new identities, some of the differences in thinking became clear in negotiations over a group of twenty-nine scientific instruments, mostly sundials. These had originally been acquired and displayed by the Art Museum, but were lent to the Educational Collections (in the Science Division) in February 1883. The loan was "for exhibition with other scientific apparatus" in the Western Galleries, which had previously housed the 1876 Special Loan Exhibition and were now being used for the Science Division's displays.[15] All twenty-nine were still to be found in the same galleries in 1912, when correspondence began as to their status.

What the discussions revealed was that these same objects were thought about quite differently by each division and its successor museum. The acquisition registers and displays of the Art Museum had concentrated primarily on decoration. Thus a sundial acquired in 1865 was described as "Gilt bronze, formed by a truncated column on a pedestal, engraved with costume figures".[16] Labelling the same object for the Western Galleries, the Science Division staff added a substantial technical description:

> On each face of the pedestal is engraved a vertical dial for the aspects N.E., S.E., S.W., N.W. The horizontal styles of these are missing, The cap of the column which carried the gnomon is also missing. Vertical lines on the columns correspond to the divisions of the zodiac. The gnomon being adjusted by means of these to the sun's position in the zodiac, the instrument would be turned until the shadow of the gnomon fell vertically in the column. The position of the point of the shadow with reference to the oblique hour lines would then indicate the time.[17]

15 "Register of Museum Property, Vol. I, 1857–1888": London, Science Museum Archive, Z46/1; John Liffen, "Behind the Scenes: Housing the Collections", in Morris, op. cit. (n. 3), pp. 273–293: 275.

16 *Inventory of the Objects*..., cit. (n. 11), museum no. 639–1865.

17 "Register of Museum Property...", cit. (n. 15).

Similar considerations were also foremost for the Science Museum staff as they decided which of the objects might be allowed to return to the V&A in 1912–1913. Items that were unlike others in the collection (*Fig. 3.4*) or had distinguishing technical features should be retained, since they could help to illustrate a taxonomy of progress. This echoed the thinking of the 1880s, when the whole group had originally been borrowed by the Educational Collections "to make more complete the collection of dials &c." Only those that failed to meet these criteria were to be given back, such as a sundial which was "not a precision dial", its interest being "more artistic than scientific".[18] By contrast, the V&A's curators wanted more of the instruments than they finally received, arguing that "many of them are finely decorated and have a considerable interest for us".[19] In the end, it was the Science Museum's arguments that took precedence: they retained twenty-five of the instruments, returning just four to the V&A.

The subsequent history of these items is worth noting. A number, including the column dial in *Fig. 3.4*, remained on display at the Science Museum and have featured prominently in publications.[20] In 2013, many were still on display in the Measuring Time gallery, having been there since its opening as the Time Measurement gallery in 1965. Those retained by the V&A have had a less glorious history, with all four currently in store and two of them designated as fakes. This reflects the generally low-key display history for scientific instruments at the V&A in the twentieth century. By the 1980s and 1990s, for example, most of those on display were being shown either as examples of decorative Islamic metalwork or in a small group of cases of European instruments tucked away in the musical instruments gallery, again primarily presented as examples of metalworking technique.[21]

A Notable Highlight

One scientific instrument, however, has been prominent in the arts displays at South Kensington ever since its acquisition. This is a clockwork globe made by Georg Roll and Johannes Reinhold in 1584 (*Pl. 3*) and originally purchased

18 Ibid.

19 "Nominal File V&A Museum": London, Science Museum Archive, file no. 1776.

20 For example, see Francis A.B. Ward, *Timekeepers*, London, 1963, nos. 2, 4, 7 and 10, all of which were among the twenty-nine instruments discussed with the V&A in 1912–13.

21 The Musical Instruments gallery closed in 2010.

FIGURE 3.4 *Column sundial, said to have been French, 17th century at the time of acquisition by the Art Museum in 1879, now identified as European, 17th century*
(SCIENCE MUSEUM 1883-132; © VICTORIA AND ALBERT MUSEUM, LONDON).

by the Holy Roman Emperor, Rudolf II.[22] This globe and its display treatment betray some of the other motives evident in the Art Museum and the V&A, in particular an appetite for objects with notable historical provenance. This had always been a feature of the art displays at South Kensington, although any historical leanings could be felt to be in tension with the Museum's educational remit (as was also the case in the Science Division).[23]

Acquired as a gift from Robert Goff in 1865, the globe was initially exhibited above the South Court, which had opened in 1862, allowing the Museum to reinstate a materials-based arrangement more forcefully. In 1868, the museum's guide described it as a "remarkable" piece, displayed in the Prince Consort Gallery, which ran above and between the two halves of the South Court (*Fig. 3.5*). Here it sat among "the most interesting and costly possessions of the Museum". It was part of a case of "Clocks and other scientific instruments, chiefly of German manufacture," placed between other cases displaying "objects in the Precious Metals generally" and "Early English work" including the twelfth-century Gloucester candlestick.[24] Within a few years, these objects were moved down into the South Court itself, with the 1885 guide noting that the globe was now displayed alongside "vessels in the precious metals designed for secular use" in a section of "Gold and Silver Work".[25] Seven years later, a guide to London noted it as one of the "objects of greatest importance" in the Western Section of the South Court, still displayed with timepieces and other instruments among cases of fine metalwork in gold and silver.[26] Although the display context was primarily about metalworking technique, the published descriptions also made much of the royal provenance. Indeed, the globe joined many other objects at South Kensington with equally notable associations: stirrups owned by Emperor Charles V, Queen Elizabeth I's virginals, a harp said to have belonged to Marie Antoinette, and items associated with the Medicis.

During the course of the twentieth century, the changing ways in which the globe has been shown have mirrored wider changes in the Museum's displays. In 1909, the opening of the Aston Webb building (now the front entrance) created more display space and heralded a major reorganization, led by a Committee on Re–arrangement, formed the year before. In considering the

---

22 John F. Hayward, "The Celestial Globes of Georg Roll and Johannes Reinhold", *The Connoisseur* 125 (1950), pp. 167–172 and 220; Hayward, "The Roll and Reinhold Celestial Globe of the Emperor Rudolph II", *The Connoisseur* 183 (1973), pp. 94–96.

23 Dunn, op. cit. (n. 10).

24 *A Guide to the South Kensington Museum*, London, 1868, pp. 36–38.

25 *A Guide to the South Kensington Museum*, London, 1885, pp. 25–27.

26 Karl Baedeker, *London and its Environs*, London, 1892, pp. 285–286.

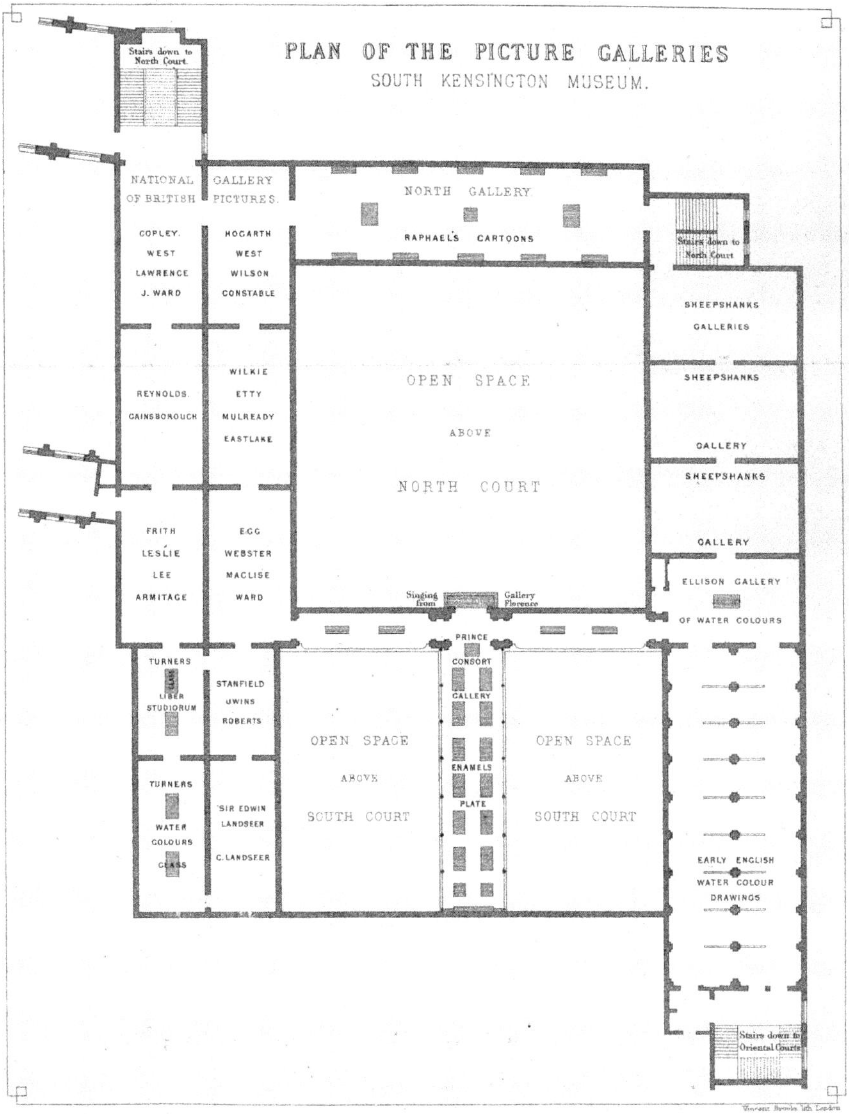

FIGURE 3.5 *Plan of the South Kensington Museum in 1868, when the mechanical globe was in the Prince Consort Gallery, above the South Court. The globe later moved into the western part of the South Court*
(PRIVATE COLLECTION).

options, the Committee faced a basic choice between a materials-based arrangement, as had been largely pursued since the Museum's foundation, and an aesthetic/historical presentation that many European museums had begun to adopt by this time, influenced in particular by Alois Riegl's *Stilfragen* (Questions of Style) of 1893. After some deliberation, however, the Committee chose to retain the materials-based approach in order to support the Museum's continued aim of stimulating artisans and manufacturers. The decision seems to have been heavily influenced by one of its more vociferous members, Lewis Day, for whom, "A museum is not a story book, but a work of reference". The materials-based approach also reflected contemporary academic studies of the decorative arts, which again were primarily by material. It was a field in which V&A curators played a significant role.[27]

Towards the end of the Second World War, however, a new Director, Leigh Ashton, moved the Museum away from a purely materials-based approach. Instead, he divided the galleries into two types: Primary Galleries containing displays by style, period or nationality; and Secondary Galleries retaining a materials-based approach.[28] As part of Ashton's redisplay, the mechanical globe moved into a primary gallery of Continental Art (*Fig. 3.6*), the 1954 museum guide recording that, "A case of small clocks and a magnificent German astronomical globe with clock, stand in the centre of the room". The same space contained furniture, tapestries, an Italian virginal, ceramics, glass and other metalwork, including a bronze relief of Rudolf II that served to reinforce the globe's royal provenance. It then remained in the same galleries for the next fifty years, though later moving to a wall case, where it was displayed in the context of 'The cabinet of curiosities'.

More recent developments have modified Ashton's distinction only slightly, designating the two types of gallery as Art and Design Galleries, which concentrate on the uses of objects within the cultures for which they were produced, and Materials and Techniques Galleries, which focus on production. In 2009, the globe moved into the Medieval and Renaissance Galleries, which, like other newer galleries at the V&A, have a greater focus on social and cultural contextualisation. Here the context is still the cabinet, as part of a display on 'Collecting: Treasures and Curiosities', which includes other instruments, clocks and precious commodities, such as shells and eggs made into vessels with ornate mounts in gold and silver. While the labels do discuss function (showing the movements of the Sun, Moon and stars, and time-telling), the

27 Burton, op. cit. (n. 3), pp. 158–164.
28 Ibid., pp. 191–199.

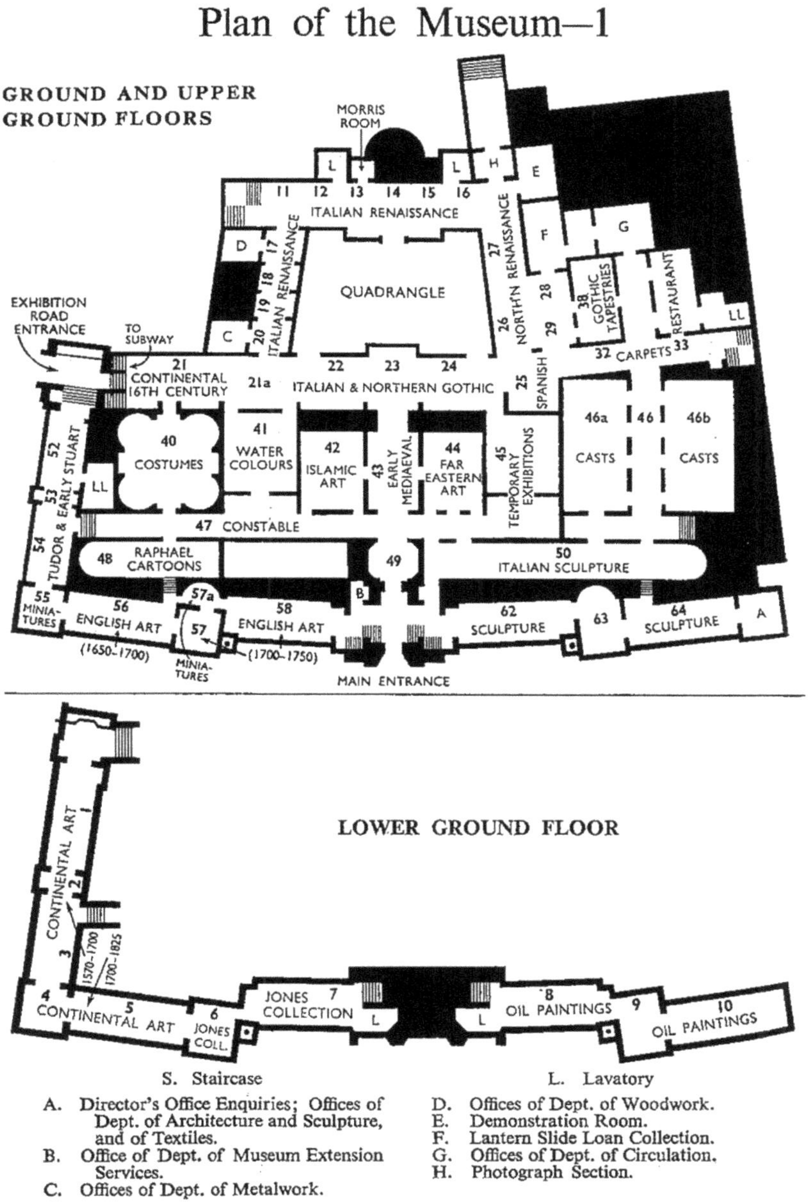

FIGURE 3.6 *Plan of the V&A in 1954. The mechanical globe was displayed in Room 1 on the lower ground floor*
(PRIVATE COLLECTION).

main message is still about Rudolf II's ownership and the conspicuous display of wealth and learning in the late-sixteenth century.[29]

The Display of Islamic Instruments

Similar reinterpretations have occurred in the display of the V&A's Islamic instruments, many of which arrived in the nineteenth century through the activities of Major Robert Murdoch Smith, Director of the Persian Telegraph Department, who was acting as an agent for the South Kensington Museum by the 1870s, purchasing objects that included astrolabes and globes.[30] It was also thanks to Murdoch that the South Kensington Museum purchased the substantial collection of Jules Richard, which included some additional instruments.

Unusually, given the materials-based emphasis at South Kensington, the Islamic material was being shown together by the end of the nineteenth century, although it was split up according to material as part of the 1908/9 rearrangement (e.g. with cases of 'Saracenic and Persian metalwork'). After the Second World War, however, Islamic material was brought together in a primary gallery, with a secondary gallery for the remaining Islamic metalwork, which is where most of the instruments were displayed until the end of the century. The displays of Islamic metalwork placed an emphasis on the fine exterior decoration, which meant that instruments such as Qibla indicators were shown closed in order to highlight the finely worked motifs on their outer surfaces.

When the primary Islamic gallery was redisplayed in 2006, however, instruments were incorporated more completely into the displays. With the gallery's emphasis on a more general cultural and social historical approach,

---

29 Contrast Michael Korey's approach to a similar globe once owned by the Elector of Saxony: "Though not a precise model of the heavens as seen from the Earth nor a research tool for the discovery of new knowledge, Reinhold and Roll's creation is much more than a purely decorative collector's item. In fact, it is rather part of a larger program linking the mechanical knowledge displayed on its surface with the political role of its princely owner." Michael Korey, *The Geometry of Power*, Munich-Berlin, 2007, p. 9.

30 Jennifer Scarce, "Travels with Telegraph and Tiles in Persia: From the Private Papers of Major-General Sir Robert Murdoch Smith", *Art and Archaeology Research Papers* 3 (1973), pp. 70–81. In a guidebook written for the Museum, Murdoch Smith noted that "some ancient astrolabes bearing the names of renowned makers [...] still exist in Persia and are valued at the most extravagant prices"; Robert Murdoch Smith, *Persian Art*, London, 1876, p. 36.

the instruments have been looked at in a new way. The display of astrolabes (*Pl. 4*), for instance, now includes a disassembled example to explain the different parts and their function. Meanwhile, the Qibla indicators, shown in a case on 'Art for Religion', are displayed open to show the functional elements in the context of Islamic religious practice.

Conclusion

Most often, decorative arts museums such as the V&A have accommodated scientific instruments into their displays by concentrating on decorative features or, more recently, social and domestic contexts. For example, the label for a barograph on display in the V&A's British Galleries until recently exemplified this neatly, informing visitors that,

> Adam-style motifs have transformed this clock and barograph from a standard long-case clock into a pure architectural form. Adam may have been involved in its making as he was designing a London house and a model village in Cumbria for its owner, Lord Lonsdale, at this period.[31]

This kind of approach is straightforward for items such as clocks, barometers, globes and armillary spheres, which were commonly made as furniture (often deploying the same artisans for stands and cabinets) with specific aesthetic styles or domestic contexts in mind. Equally, one can situate historic sundials and other pocket instruments as primarily personal accessories or luxury commodities. Indeed, one could go so far as to argue that many of the historic instruments that survive in museums, including those with 'scientific' collections, come predominantly from this same context.[32]

It is important to recognize the value and legitimacy of such approaches, which encourage alternative, yet historically valid, interpretations of scientific instruments of the sorts preserved in museums of all sorts. In the recently redisplayed Base Metals Gallery (a 'Materials and Techniques' gallery), for example, the V&A has been free to display scientific instruments in contexts where an instrument historian's concerns over the question, 'What is a

31 Label for a barograph clock with a movement by Alexander Cumming, about 1774 (on loan from the Fiorencia Art Foundation), which was still on display at the V&A in 2010.

32 Field, op. cit. (n. 1); see also Dunn, "Made to Measure: Some Thoughts on the Design of Scientific Instruments", in Liba Taub, Frances Wilmoth (eds.), *The Whipple Museum of the History of Science. Instruments and Interpretations*, Cambridge, 2006, pp. 121–137.

scientific instrument?', become at best secondary and, in practice, unimportant. In this gallery, the curators acknowledge some of the technical and functional features in which a museum of science might be more interested, yet remain focused on issues of interest to a decorative arts museum. Perhaps, then, looking at how other types of institution think about the scientific instruments they display can offer some hints to instrument historians as they seek to think creatively about the different ways in which their objects of choice can be examined, interpreted and displayed so as to excite the interest of others.

CHAPTER 4

"Of Sufficient Interest..., but not of Such Value...": 260 Years of Displaying Scientific Instruments in the British Museum

Silke Ackermann

Introduction

Since its founding in 1753, the British Museum in London has had a collection of mathematical and optical instruments. The justification for their presence has often been questioned and the rationale for their display—both in permanent and in temporary exhibitions—has changed considerably over the years.

This essay is intended as a first overview of this story from the Museum's opening in 1759 to the present day, 260 years later. This story can roughly be divided into six major phases, here recounted based mostly on contemporary guide-books.

First Phase: Eighteenth Century—The Beginning

The first official description of the British Museum after its opening to the public in 1759 can be found in Edmund Powlett's guidebook of 1761, offering a fascinating insight into displays at the Museum's first home at Montagu House.[1] It is here, on our tour of the First Floor, that we encounter the instruments—after having moved through the first room, the saloon, the five rooms of the

1 Edmund Powlett, *The General Contents of the British Museum: With Remarks Serving as a Directory In Viewing that Noble Cabinet*, London, 1761; *Second Edition, With Additions and Improvements, and a Complete Index*, London 1762. The "ancient mathematical instruments" here referred to include all or at least most of the fifty-seven listed as "Mathematicall Instruments & c." in the inventories of Sir Hans Sloane, the acquisition of which led to the founding of the British Museum in 1753. Cf. Silke Ackermann, Jane Wess, "Between Antiquarianism and Experiment: Hans Sloane, George III and Collecting Science", in Kim Sloan (ed.), *Enlightenment: Discovering the World in the Eighteenth Century*, London, 2003, pp. 150–157 and 289: 152–154; John Cherry, "Medieval and Later Antiquities: Sir Hans Sloane and the Collecting of History", in Arthur McGregor (ed.), *Sir Hans Sloane: Collector, Scientist, Antiquary—Founding Father of the British Museum*, London, 194, pp. 198–221: 213–215.

 | DOI 10.1163/9789004264403_005

Department of Manuscripts, Medals and Coins, into the Department of Natural and Artificial Productions:[2]

> In the other Repository near the Windows are a great Number and Variety of ancient mathematical Instruments, by which the learned Observer may be enabled to judge how much that particular Branch of Science is improved.[3]

But this is just one of two opportunities to come across instruments in the building. Having moved back downstairs into the Department of Printed Books on the Ground Floor and the Reading Room:

> We have one Room more to mention, which contains some Sea Compasses, improved by Dr. Knight, such as are now used in the royal Navy, and several Magnets, and Apparatuses, serving to shew the magnetical Powers in philosophical Uses.[4]

That the Museum was in some way actively involved in the sciences and contemporary instruments was in no way surprising to visitors of the time. The changing expectations of what constitutes science and how it should be represented in a collection in the late eighteenth century is well illustrated by a letter supposedly written by Matthew Bramble, a well-to-do landowner from Wales and one of the main protagonists in Tobias Smollett's *The Expedition of Humphry Clinker*, first published in 1771. The recipient, Bramble's friend, a Dr Lewis, is informed about the family's travels through England and Scotland, during the course of which they visit London:

> Yes, Doctor, I have seen the British Museum; which is a noble collection, and even stupendous, if we consider that it was made by a private man, a physician, who was obliged to make his own fortune at the same time [...]. I could wish [...] the whole of the animal, vegetable and mineral kingdoms completed, by adding to each, at the public expense, those articles that are wanting. [...] I could also wish, for the honour of the nation, that there was a complete apparatus for the course of

---

2 Powlett, op. cit. (n. 1), p. 34 ff.

3 Ibid., p. 58.

4 Ibid., pp. 209–210. For Gowin Knight, the first 'Principal Librarian' (i.e. director) of the British Museum and his interest in science, especially magnetism, see: David M. Wilson, *The British Museum: A History*, London, 2002, pp. 28 and 39, plate 7.

> mathematics, mechanics, and experimental philosophy; and a good salary settled upon an able professor, who should give regular lectures on these subjects.[5]

'Ancient' science being juxtaposed with contemporary inventions, to be looked at and to be enlightened by, aided by lectures on the relevant subjects—science played, and was expected to play, a natural role in the display of all branches of knowledge displayed at the Nation's 'Noble Cabinet' in the first decades after its opening.

Second Phase: First half of the Nineteenth Century—Stagnation

From being a place in the late eighteenth century that Hans Sloane might still have recognized with its displays of manuscripts, printed books, coins and medals, and 'natural and artificial productions', the Museum changed rapidly during the first half of the nineteenth century. This is best demonstrated through the acquisition of classical antiquities that are still amongst its most iconic objects. The Assyrian Winged Bull, the Parthenon Sculptures, and the Rosetta Stone, to name but a few,[6] all entered the Museum at this time and the changing departmental structures reflect this changing identity.[7] Montagu House soon became much too small to hold the rapidly growing collections. The 'King's Library', completed in 1827 as the first wing of the new Museum building that we know today, was prompted by the gift of King George III's (reigned 1760–1820) private library (rather confusingly also called 'The King's Library') to the nation by his son King George IV (Prince Regent from 1811, reigned 1820–1830) in 1823.[8] Interestingly, the King's famed collection of scientific instruments, kept at Kew Observatory in Surrey until the Government decided to discontinue that building's maintenance, was not subsequently given to the

5 This is the opening paragraph in the letter dated 2 June: Tobias Smollett, *The Expedition of Humphry Clinker*, London, 1985, p. 133. For a brief overview of collecting science in the Enlightenment see: Ackermann, Wess, op. cit. (n. 1).

6 See respectively: British Museum (BM), Department of the Middle East, reg. no. ME 118872 www.britishmuseum.org/explore/highlights/highlight_objects/me/c/colossal_statue_winged_bull.aspx; Department of Greek and Roman Antiquities www.britishmuseum.org/explore/highlights/article_index/l/lord_elgin_and_the_parthenon_s.aspx; and Department of Ancient Egypt and Sudan, reg. no. EA 24, www.britishmuseum.org/explore/highlights/highlight_objects/aes/t/the_rosetta_stone.aspx (accessed 22 February 2014).

7 For a concise overview over these changes see: Wilson, op. cit. (n. 4), p. 379.

8 Cf. Ibid., pp. 78–79 and 99.

British Museum. Instead the collection was donated to King's College London in 1841 where Prince Albert opened the King George III Museum in 1843.[9]

'Science', which for visitors of the eighteenth century had been a matter of discourse and material expected to be on display in the Museum, had all but vanished from view—at least as far the official guide books are concerned. Mentions of instruments are few and far between.[10] In the 26th edition of the *Synopsis* in 1832, only one instrument is mentioned as being on display; in Room VI (Greek and Roman Sculptures) of the Department of Antiquities, as "no. 58*. A sun-dial. Purchased in 1821".[11] In 1848 we find "a Roman sundial in the Elgin Room",[12] and in the "Ethnographical Room [...] in cases 1–2", amongst objects from China, "Chinese mariners' compasses" and a "viatorium, consisting of a compass and small portable sun-dial. Presented by Albert Way, Esq.".[13]

It is only from the middle of the 19th century onwards that matters change once again:

> The medieval Collection [...] is arranged partly with reference to the material of which the objects are formed, partly to the purpose for which they were intended. A large portion of the Collection is about to be rearranged in consequence of the extensive additions that have been made to it.[14]

---

9 For the King George III Museum see the online archive catalogue of King's College London: www.kingscollections.org/catalogues/kclca/collection/g/10ge50-1/ (accessed 22 February 2014). The instruments which had been given for the purpose of "teaching a general course in experimental philosophy", remained at King's College until the closure of the Museum in 1926, when they were deposited on loan at the recently established Science Museum in South Kensington. On George III's collection now at the Science Museum, see: Alan Morton, Jane Wess, *Public and private science: the King George III collection*, Oxford, 1993.

10 Cf. for example the various editions of the *Synopsis of the contents of the British Museum*, published in London in the years 1808 to 1856.

11 *Synopsis of the contents of the British Museum*, 26th ed., London, 1832, p. 197. The sundial in question can be identified as BM reg. no. 1821,0301.1 (PRN GAA83078).

12 *Synopsis or Guide Book to the British Museum, Part IV: Greek, Roman and Miscellaneous Antiquities*, London, 1848, p. 37, no. 186.

13 *Synopsis or Guide Book to the British Museum, Part V: Artificial Curiosities*, London, 1848, p. 4. The instrument presented in 1846 by Albert Way, the founder of the Archaeological Institute in 1845 and friend of Augustus Wollaston Franks (for him see below), can be identified as BM reg. no. 1846,0616.1 (PRN EAS34006).

14 *Synopsis of the contents of the British Museum*, 63rd edition, London 1856, pp. 265–266.

These re-arrangements and additions, as well as the rationale behind them, made a fundamental difference to the display of instruments, and they derived from the efforts of one man, Augustus Wollaston Franks.

Third Phase: Late Nineteenth Century—The Franks Years

In the second half of the 19th century the British Museum developed into the institution that today's visitors would recognize. In 1880, after lengthy discussions about the lack of space, the natural history collections, that had been one of the major attractions in its early history, were finally moved off-site into a new building in South Kensington that later became the (independent) Natural History Museum.[15] At the same time, in 1879, most of the portraits were transferred to the National Portrait Gallery near Trafalgar Square that had been founded in 1856.[16]

It is during this period that a man enters the scene whose influence on the development and display of the instrument collections can hardly be exaggerated. Augustus Wollaston Franks (1826–1897), Keeper of the newly formed Department of British and Medieval Antiquities (1851–1897), expanded the collections beyond printed material and classical antiquities. Franks, a Gentleman who accepted the post at the British Museum out of interest, not necessity, could arguably be called one of the greatest collectors of his time as well as a friend of many great collectors. He was thus able to bring into the Museum a great many objects not only through careful purchases, but also through systematic steering of large donations and bequests, and by careful manoeuvring among the relevant stakeholders in the Museum. Franks focussed on hitherto undervalued ethnographic and archaeological material as well as art objects from Europe and the Orient. "He so altered its collection policies and departmental structure that he is regarded by many as the founder of much of today's British Museum".[17]

15 Wilson, op. cit. (n. 4), pp. 142 and 184.

16 Ibid., pp. 9, 29 and 178.

17 Marjorie Caygill, "Franks and the British Museum—The Cuckoo in the Nest", in M. Caygill, J. Cherry (eds.), *A.W. Franks. Nineteenth-Century Collecting and the British Museum*, London, 1997; pp. 51–114: 52. See also: D.M. Wilson, "Augustus Wollaston Franks—Towards a Portrait", in Ibid., pp. 1–5, and August W. Franks, "Apology of my Life", in Ibid., pp. 318–331.

This general assessment also holds true for the scientific instruments in particular.[18] At the time no specialized museum of science existed anywhere in Britain, nobody seriously collected the instruments although many had survived largely forgotten in cabinets, societies and laboratories.[19] Franks and some of his friends were the exception and those instruments that ultimately came into the British Museum via various routes (acquisitions, gifts and finally bequests by Franks himself as well as gifts and bequests by his friends) formed "the largest and most impressive museum collection of early instruments".[20] This only changed "when, in 1922, Lewis Evans (1853–1930) offered [...] his collection to the University of Oxford".[21] Nowadays, this dominance of the British Museum in terms of scientific instrument collecting may strike us as odd since we are used to specialised museums dedicated to the subject. It is thus interesting to note that the uncertainty regarding this material continued for much of the second half of the nineteenth century. It was only with the large loan exhibition of scientific instruments in 1876, held at the Museums in South Kensington, that this situation changed; finally in 1909 an independent Science Museum was established.[22]

Franks did not only collect and encourage the collection of instruments;[23] he also radically changed the displays. After being all but absent in the earlier guide books, from the late 1850s onwards we find on the upper floor as Room 25[24] the new 'British and Mediaeval Room' now fully arranged, with cases 104 and 105 described as "Horodeictical [sic.] Instruments,[25] consisting of astrolabes, dials, and watches of various kinds".[26] By 1866 this is much

18 Cf. Robert G.W. Anderson, "Early Scientific Instruments", in Ibid., pp. 286–295, and Rachel Ward, "'Islamism, not an easy matter'", in Ibid., pp. 272–285.

19 Anderson, op. cit. (n. 18), p. 293. See also: Wilson, op. cit. (n. 4), pp. 176–177.

20 Anderson, op. cit. (n. 18), p. 293.

21 Ibid., p. 294.

22 Cf. Wilson, op. cit. (n. 4), pp. 176–177. On the Science Museum see the essay by Alison Boyle, *infra*, pp. 37–60.

23 As well as clocks and watches, the latter being a particular interest of his, cf. Wilson, op. cit. (n. 4), p. 177.

24 In the 2014 numbering of the British Museum rooms this is 69.

25 The origin of the unusual term 'horodeictical' remains unclear. The *Oxford English Dictionary* does not list the word. It is possible that it may derive from "horodix"—from Greek ωρα (= hour) plus δειξις (= exhibition), from δεικ- (= to show)—used in two sources in the late seventeenth and early eighteenth centuries, referring to an instrument that shows the passing of hours; cf. *Oxford English Dictionary*, s.v. "horodix".

26 *British Museum. Guide to the Exhibition Rooms of the Departments of Natural History and Antiquities*, London, 1859, p. 101. The identical layout is described in the *Guide* of 1862.

expanded; the sections of the previously anonymous *Guide* describing the 'British and Medieval Room' (still room 25) and the adjacent 'Ethnographical Room' (the adjacent Room 26)[27] are for the first time signed by Franks himself. The objects are described in greater detail and we find instruments now in Cases 108–115 amongst "Metal Work of various kinds; ecclesiastical relics [. . .]; a collection of horodeictical instruments, such as astrolabes, quadrants, and dials of various kinds; vases and dishes; arms and armour".[28]

Shortly afterwards, in 1867, the display has changed again. Franks is clearly working on constantly improving and adjusting the galleries following various spectacular acquisitions. The instruments are now no longer displayed in cases 108–115; instead we find that "in Upright Case M is a curious piece of clockwork in the form of a ship, presented by Octavius Morgan, Esq., M.P., and a collection of horodeictical instruments, such as astrolabes, quadrants, and dials of various kinds".[29]

With the removal of the natural history collections to South Kensington in the 1880s, "the idea of the British Museum as a universal museum was preached for the first time".[30] The new extensive *Guide to the Exhibition Galleries of the British Museum, Bloomsbury*—published in frequently updated editions between 1879 and 1912—reflects in greater detail the changes to the displays on the main site resulting from a further 5,713 square metres having become available for the antiquities departments.[31] The Indian collections are now far more prominent following the transfer of some of the objects (including the famous Amaravati sculptures) from the India Museum in 1879.[32] It is at this point that we find a plea from the Principal Librarian Edward A. Bond that the collections may be used for academic teaching.[33] By 1886 the layout and numbering of the galleries has changed completely. The medieval collections are now on display in the new room 41—in exactly the same place where a visitor

27 In the 2014 numbering of the British Museum rooms this is 68.

28 *British Museum. A Guide to the Exhibition Rooms of the Departments of Natural History and Antiquites*, London, 1866, p. 114. Interestingly, watches are not mentioned.

29 *British Museum. A Guide to the Exhibition Rooms of the Departments of Natural History and Antiquities*, London, 1867, p. 114. The same description can be found in the *Guide* of the years 1869, 1871 and 1877. The "curious piece of clockwork in the form of a ship" is the automated galleon made by Hans Schlottheim in c. 1585, BM reg. no. PE 1866,10–30,1.

30 Wilson, op. cit. (n. 4), p. 184. The new building was formally opened in 1881, with the last objects from the zoological collections finally having moved in 1883.

31 Ibid., p. 184.

32 Ibid., p. 172.

33 *Guide to the Exhibition Galleries of the British Museum, Bloomsbury*, London, 1886, p. xxxii.

FIGURE 4.1 *The British and Medieval Room in the British Museum circa 1875. The upright case with the 'Nef' and the astrolabes is on the right-hand side of the photo* (BM CENTRAL ARCHIVE, YORK ALBUM 119; © TRUSTEES OF THE BRITISH MUSEUM).

in 2013 would also find them (*Fig. 4.1*).[34] The objects are largely organized by type, such as metalwork, astrolabes, Limoges enamels, ivory carvings etc. This includes, in cases 7 to 10, metalwork from the Islamic World.[35] In cases 15 and 16 we find:

34 In the 2014 numbering of the rooms this is 40.

35 Ibid., p. 204.

> Astrolabes and Clocks. Among the former is one made for the Sultan of Damascus, in 1235;[36] another for London about 1290,[37] and an English specimen made by Blakeney, 1342.[38] Two of the clocks presented by Octavius Morgan, Esq., deserve notice; one is in the form of a ship, and was made for Emperor Rudolph II (1576–1612); the other was made in Cracow by Lucas Weydman.[39]

Table Case A contains:

> Historical Relics. [. . .] hourglass of Stephen Bathori, king of Poland;[40] dial of Robert Devereux, Earl of Essex;[41] [. . .] quadrant of Richard II;[42] astrolabe of Henry VIII,[43] quadrant of Edward VI;[44] astrolabe of Henry, Prince of Wales, son of James I,[45]

whilst in Table Case K we find:

> A collection of watches, chiefly bequeathed by Lady Fellows; a series of sundials of various periods.[46]

It is intriguing to note that similar display criteria were still at work when the present author joined the British Museum in 1995.

Only four years later, in 1890, the collections of instruments, and clocks and watches has been much expanded, with these changes clearly visible in the displays. Rather than the previous two cases (15 and 16) there are now four cases (13 to 16) dedicated to this material, introduced with "Astrolabes and Clocks, mostly given or bequeathed by Octavius Morgan, Esq. with the description of

36 BM reg. no. 1855,0709.1.

37 The so-called 'Sloane astrolabe', reg. no. SLMathInstr.54.

38 BM reg. no. 1853,1104.1

39 *Guide to the Exhibition Galleries of the British Museum, Bloomsbury*, London, 1886, p. 204.

40 BM reg. no. PE 1872,0308.1 (PRN MCC6669).

41 BM reg. no. PE 1866,0221.1 (PRN MCC3846).

42 BM reg. no. PE 1860,0519.1 (PRN MCC3467).

43 BM reg. no. PE 1878,1101.113 (PRN MCC3715).

44 BM reg. no. PE 1858,0821.1 (PRN MCC3479).

45 BM reg. no. PE 1855,1201.223 (PRN MCC3783). *Guide to the Exhibition Galleries of the British Museum, Bloomsbury*, London, 1886, p. 205.

46 Ibid., p. 206.

the contents continuing as before".[47] In addition to the table cases A and K, table case G is now used to show

> Watches and sundials, bequeathed by Octavius Morgan, Esq. The watches are arranged in order of date, and following the plan adopted by Mr. Morgan. They range from the early years of the 16th century to the beginning of the present century. The sundials embrace about the same period.[48]

Fourth Phase: The First Half of the Twentieth Century—"The Instruments Belong into the Science Museum"

Until the First World War the Museum saw few developments in the display of instruments as witnessed by the guidebooks. The numbering of the galleries occasionally changes due to new collections coming in. European and Oriental material remains together. What is now explicitly stated, however, is the role that the astronomical instruments are supposed to play in the displays, namely to show "very fully the progress of the manufacture of time-pieces from an early date".[49] This attitude, that instruments were merely one step in a logical and linear story continuing from the earliest sundials to the latest development in clock-making, influenced the way the instruments were regarded for many decades to come. That different cultures or societies might require different ways of measuring time for different reasons, resulting in supposedly different 'stages' of the development occurring synchronously, is a concept that did not enter the British Museum displays until much later.

It is only with the opening of a new wing, the Edward VII galleries on the north side of the Museum in 1914, that major changes to subject divisions and object displays occur. The 1914 to 1925 summary guides allow us, albeit in very

---

47 *Guide to the Exhibition Galleries of the British Museum, Bloomsbury*, London, 1890, p. 237. The Morgan bequest dates from 1888.

48 Ibid., p. 237. Charles Octavius Swinnerton Morgan (1803–1888), Conservative MP for the county of Monmouth (1840–1874), was a noted collector and antiquary, and a personal friend of Franks'. He bequeathed his collection of clocks, astronomical instruments and papal rings to the British Museum. Cf. David Thompson, "Octavius Morgan, Horological Collector", *Antiquarian Horology* 27 (2003), pp. 302–314, 406–422 and 618–642; *Antiquarian Horology* 28 (2004–2005), pp. 337–347 and 602–618; *Antiquarian Horology* 29 (2005–2006), pp. 189–216 and 629–644.

49 *Guide to the Exhibition Galleries of the British Museum, Bloomsbury*, London, 1907, p. 88.

brief descriptions, to follow these changes step by step, year after year, until in 1925 the new arrangement of the 'British & Medieval' material on the ground floor is completed.[50] We now find in the "Western half, European pottery and porcelain [...]. Watches and dials", and in the "Eastern half, Oriental pottery and porcelain and the collection of Glass. At the east end is the Waddesdon Room, [...] The western alcove is arranged as Franks Room, containing a portion of the bequest of Sir A. Wollaston Franks, including especially the Royal Gold Cup".[51]

In the summer of 1931, Ernst Zinner (1886–1970), the distinguished German historian of Renaissance astronomy and author of many publications on scientific instruments, attended the *Second International Congress of the History of Science and Technology* in London. Zinner used any occasion that presented itself on his travels to visit collections that might be relevant for his research on manuscripts and instruments and noted his observations quite candidly in his personal diaries. It is thanks to these diaries, now preserved in the Archive of Frankfurt University, that we possess one of the rare first-hand records of instruments on display in the British Museum galleries shortly before the Second World War:

> The British Museum has nice collections of manuscripts and antiquities. But the ethnographical collections on the 1st floor are a mess and much too crowded. The beautiful Rothschild collection should be in the Victoria and Albert Museum, the instruments belong into the Science Museum. There should be a Museum of Mankind and the Antique, Babylonian and Egyptian things could nicely be distributed. But after all that there would still be no room for British culture—this is completely missing. In fact, how the English used to live is not illustrated anywhere in London.[52]

According to Zinner, the British Museum clearly had rather failed in its aspiration to be a 'universal museum', at least as far as the displays were concerned.

50 Room 33 in the 2014 numbering of the galleries.

51 *Summary Guide to the Exhibition Galleries of the British Museum*, London 1925.

52 Zinner-Archive, Goethe-Universität, Frankfurt am Main, S Zi 25, I Zi 156: Reisen 1894–1943, "Reise nach England, 14. Juni–15. Juli 1931", p. 19 (translation from the German original by the present author). For information on the Zinner-Archive see: Ackermann, "Dormant Treasures: The Zinner-Archive at Frankfurt University", *Nuncius* 16 (2001), pp. 711–722.

A good two decades later, however, Zinner found the displays much changed. In September–October 1954 he travelled to France to attend the *International Congress of Chronometry* in Paris, from where he continued to England to participate in a meeting of the Horological Society in the Science Museum. Once again he paid a visit to the British Museum and noted:

> Then we visited the various departments, whereby I paid particular attention to the scientific instruments. An assistant of Peter Laslo's, who is in charge of medieval and modern material,[53] facilitated my examination of the instruments in the cupboards underneath the showcases. [...] There are far fewer instruments on display than in 1931 because the Edward VII gallery now also has to house Prehistory, the Ancient World, Ethnography etc. Consequently, there is only a small selection of each subject on show [...]. A check of the hand-list in Lasko's room shows that there are far fewer instruments listed in the hand-list than there are in the old, very detailed registers and my own notes.[54]

There is no evidence in the diary that Zinner knew what had happened between his two visits or why some of the instruments that he earlier had seen (or at least noted down according to the old registers) could no longer be located. It is through other sources that we gain an intriguing insight into the rationale that lay behind the display of instruments during the war years.

Even before the declaration of war on Germany in September 1939 huge efforts had been undertaken in the British Museum to remove the portable objects in the precious collections to locations less exposed. However, to keep spirits up in the face of the enemy, a special display from across the collections was arranged in 1940, the so-called "War exhibition",[55] for which special permission was sought from the Trustees (*Fig. 4.2*). The objects in this exhibition were regarded as being

53 This is Peter Lasko (the 'Laslo' clearly a typo in Zinner's manuscript), Assistant Keeper in the Department of British and Medieval Antiquities 1950–65, cf. Wilson, op. cit. (n. 4), p. 387.

54 Zinner Archive, Goethe-Universität Frankfurt am Main, S Zi 25, II Zi 156: Reisen 1946–1955, "Reise nach Frankreich und England, 20. Sept.–24. Okt. 1954", pp. 18–19 (translation from the German original by the present author).

55 Also referred to as the 'suicide' or 'sacrificial exhibition'.

FIGURE 4.2 *The "War Exhibition" in the Central Saloon 1940–1941* (BM CENTRAL ARCHIVE, WORLD WAR II IMAGES; © TRUSTEES OF THE BRITISH MUSEUM).

> of sufficient interest to justify the opening of the gallery to the public, but not of such value that damage to or destruction [of the objects] [...] would seriously affect the strength of the main collections.[56]

"Objects [...] of sufficient interest [...] but not of such value" included for the decision makers what we would now regard as some of the finest instruments in the collection at the time. A rare late medieval horary quadrant,[57] a virtually

56 S. Ackermann, J. Cherry, "Richard II, John Holland and Three Medieval Quadrants", *Annals of Science* 56 (1999), pp. 3–23: 4, n. 6. For an overview of the British Museum during the war see: Wilson, op. cit. (n. 4), pp. 249–251. "[...] a 'sacrificial' exhibition of duplicates and folk-life material from the Department of British and Mediaeval Antiquities was put on in the Central Saloon at the top of the main staircase. [...] It was the incendiary bombs [...] that caused the worst trouble [...] lodged between the copper roofs and the ceilings of a number of the galleries. Those burnt out where [...] the Central Saloon [...]. The temporary exhibition was completely destroyed", Ibid., pp. 250–251.

57 Dated to 1400, BM reg. no. PE 1856,0627.155 (PRN MCC3477).

FIGURE 4.3 *The Central Saloon after the air raids on 10 May 1941* (BM CENTRAL ARCHIVE, WORLD WAR II IMAGES; © TRUSTEES OF THE BRITISH MUSEUM).

unique astrolabe with Hebrew markings of the mid-fourteenth century,[58] and a sundial in the shape of a boat, a so-called '*navicula de Venetiis*' signed by the famous Flemish maker Arsenius and dated 1581[59] were among the objects on the list.[60] Attitudes had clearly changed since Franks had carefully built up the collection in the second half of the nineteenth century; the instruments now seemed dispensable.

On 10 May 1941 disaster struck. During air-raids the so-called Central Saloon and the adjacent rooms, where the exhibition had been located, suffered a major hit (*Fig. 4.3*). Sixteen instruments were completely destroyed or went

58 Dated to c. 1350, BM reg. no. PE 1893,0616.3 (PRN MCC 3708).

59 BM reg. no. 1893,0616.8 (PRN MCC3596).

60 For a list of the objects that were included see 1940 War Exhibition Register and Notes: British Museum, Central Archives, 'Scientific Instruments'. See also: Ackermann, Cherry, op. cit. (n. 59), p. 4, n. 6. The Hebrew astrolabe was withdrawn at the last minute as the annotations show.

missing, including the above-mentioned *navicula*. Miraculously the quadrant survived, albeit with serious fire damage.[61]

Fifth Phase: The Post-War Years—Picking Up Again

Guidebooks from the late 1950s and 1960s document very little change in the displays and the underlying attitude towards instruments, with war damage still affecting some of the galleries. It is only in the early 1970s that yet another major shift occurred that was gradually carried out in the following years. The King Edward VII building was now completely dedicated to the Departments of Prints & Drawings and Oriental Antiquities, whilst the Department of Medieval and Later Antiquities and its associated displays returned to the south-east corner of the upper floor in the main building where they can still be found today. For the first time there was a specialized Horological display (Room 44) that was dedicated to the evolution of clockwork following the acquisition of the Ilbert collection in 1958 and that remained largely unchanged until the installation of a much updated, but similarly focused gallery in 2008 (Rooms 38–39).

The instruments largely continued to play the same role in displays as they had done more or less since Franks' times: as examples of metalwork or craftsmanship, as various forms of early timekeepers before the invention of clocks without further historic contextualization or as objects illustrating 'secular life' or to highlight certain historic personalities.[62]

The Sixth Phase: The New Millennium

In the mid-1990s a major shift took place regarding the role of science and the history of science in the British Museum, gradually manifesting itself in an increased, often surprising visibility of instruments in temporary as well as permanent displays.[63] This sea change corresponded with a new focus on

61 Cf. note 56 above.

62 See for example the *British Museum Guide*, London, 1976, pp. 116, 117–118 and 134.

63 This development started with the appointment of the historian of science Robert G.W. Anderson as Director and the subsequent appointment of the first specialized curator for European and Islamic history of science, the present author. However, these appointments coincided with a sea change in the Museum as a whole and the changed displays would not have been possible without the interest and openness of many colleagues around the Museum.

FIGURE 4.4 *Display "Courtly Culture" in Room 40 at the British Museum* (© TRUSTEES OF THE BRITISH MUSEUM).

narratives in the galleries and an increasing consideration of general audiences and their needs and interests. Instruments began to be placed into a cultural and social context rather than, to name but one example, into a straight line of a perceived human progress in time-keeping. The fundamental changes in Tudor England following the dissolution of the monasteries (Room 46), transfer of knowledge between the Islamic World and Europe in the Middle Ages, intellectual discourse at the court of Richard II (Room 40: *Pl. 5 and Fig. 4.4*) or the five pillars of Islam (Room 34), to name but a few examples, are all being illustrated with instruments that naturally take their place next to other objects as would have been the case at the time of their production and use. Visitors explored the after-life of Babylon in Renaissance Europe, as discussed in the travelling exhibition *Babylon: Myth and reality*,[64] and learned about the importance of finding direction on Pilgrimage in the recent major exhibition *Hajj: Journey to the Heart of Islam*.[65] In each case instruments were not primarily deployed to explain complicated scientific concepts that even at the

64 Shown in Paris, Berlin and at the British Museum in 2008 and 2009.

65 Shown at the British Museum in 2012 and travelling at the time of writing.

time of their making would only have been accessible to very few people, but rather to highlight the prevalent questions and the ways that various peoples, cultures and societies found to answer those questions.

Conclusion

Instruments have played a number of roles in the British Museum displays, from illustrating the 'improvement of science' in the late eighteenth century to illustrating craftsmanship and the endeavour of (European) mankind to steadily improve the accuracy of timekeeping in the nineteenth and twentieth centuries to artefacts that contribute to a cultural and social narrative in their respective contexts in the early twenty-first century. From an instrumental point of view the British Museum displays have thus become truly that of a 'universal museum'—and the instruments are now not only regarded to be 'of sufficient interest' but rather of great 'value' to illustrate manifold aspects of our past.

CHAPTER 5

Instruments on Display at the Paris Observatory

Laurence Bobis and Suzanne Débarbat

Introduction

The first meeting of the Académie Royale des Sciences, created by King Louis XIV (the 'Sun-King'), took place on 22 December 1666. A few months later, on 21 June 1667, the day of the summer solstice, the building of the Observatoire Royal was decided and the reference meridian for longitudes was determined to be the axis of the future building. For the next one and a half centuries, the new institution would be supervised by the Académie Royale des Sciences.

The Observatoire Royal was erected between 1667 and 1672, but already from 1670 onwards the ground floor was accessible and in 1671 the first floor was completed, so that members of the Académie would be living in the observatory's apartments. Claude Antoine Couplet (1642–1722), the *astronome-concierge*, and Gian Domenico Cassini (or Jean-Dominique, 1625–1712) were invited by Louis XIV to join the Académie. Naturally, this new observatory attracted a number of important people of the time, and both the king of France, Louis XIV, and James II of England came on ceremonial visits.

But the first person at the Paris Observatory to think about welcoming the public and exhibiting instruments was undoubtedly Jean-Dominique Cassini (Cassini IV, 1748–1825), the fourth in his family to be associated with the institution. More than two centuries later, the Observatoire still offers the public an opportunity to view its patrimonial collections.

Cassini IV and the 'Decoration' of the Paris Observatory

Jean-Dominique Cassini was appointed director of the Observatoire Royal in 1784, at the death of his father, César-François (Cassini III, 1714–1784), otherwise known as Cassini de Thury, who had served as director since 1771. His ancestors, Gian Domenico (Cassini I) and Jacques (Cassini II, 1677–1756), had never held that post. At the time of Cassini I and II, the observatory was in good repair and the instruments were new. However, by the mid-eighteenth century, some of the small buildings that housed instruments on the eastern side of the grounds needed restoring. Cassini III, who was busy elsewhere in France

 | DOI 10.1163/9789004264403_006

(surveying), was not very interested in maintaining the site. At the end of his life, he suffered from ill health and was often absent from the observatory.

This situation may explain why in 1774 Cassini IV reported the necessity of restoration to the king of France. In 1774 Louis XV had died and his grandson Louis XVI had taken the throne. Fortunately for Cassini IV, after several requests, two ministers of Louis XVI, Breteuil and d'Angivillers, who was Directeur General des Bâtiments (General Director of Buildings), submitted their evaluation. This report was later published with some other essays about the history of the observatory.[1] Count d'Angivillers, described by Cassini IV as a "passionate friend of science, scientists, and especially my friend",[2] seems to have taken a prominent role in this project. The opinion of the Academy of Science was, of course, required. After many debates, the project was finally accepted and, in 1775, d'Angivillers approved the restoration as soon as funding was available. Renovations did not begin, however, until the spring of 1786.

The *Mémoires pour servir à l'histoire des sciences et à celle de l'Observatoire Royal de Paris* include a second treatise concerning the restoration and Cassini's views on how to renovate the building to fulfil contemporary requirements for the observatory. In this very important tract, Cassini IV discusses what he calls his "meridian hall decoration project".[3] Today we would call this a "museum project". Cassini IV was indeed clearly interested in creating a space for welcoming visitors that would also keep outsiders away from the buildings and the offices used by researchers:

> For a long time I had wanted to find a way of avoiding inopportune visits by curiosity-seekers and strangers. It had been the usual practise to allow such visits to the observatory quarters. Sometimes these visitors unthinkingly touched the instruments and moved or otherwise disturbed them despite the surveillance recommended to those responsible for guests. During the building renovation, I had the idea of assembling sufficient objects in the large meridian room to satisfy public curiosity, and from thereon, we would no longer be vexed by these visits.[4]

1 Jean-Dominique Cassini, "Rapport fait à l'Académie royale des Sciences, le 4 août 1784, sur le Projet d'établissement à l'Observatoire", in *Mémoires pour servir à l'histoire des sciences et à celle de l'Observatoire de Paris*, Paris, 1810, pp. 162–164.

2 "Premiere Mémoire", ibid., p. 10.

3 "Deuxième Mémoire", ibid., p. 53.

4 Ibid., pp. 53–54 (translated by Susan Keyes).

The fear conservators have today that old instruments and objects may be mishandled and damaged because of insufficient supervision was hardly a concern of Cassini IV. He envisioned the public freely moving among the collections! However, displaying instruments was not his only motivation. They were not only for show, but regularly used as part of the practical astronomy class that he had given at the observatory for two years. Instruments taken out of their scientific context would be without their quintessence, would be put in inexperienced hands, so to speak. Thus, after having identified the problem, Cassini reconciled museum exhibitions and public education without much thought for conservation.

Cassini decided to set up the exhibition in the most imposing room in the Perrault building, the room on the second floor traversed by the incised meridian line. He suggested a museum-type installation, using the huge pillars, 16 meters high, to partition the exhibit into four areas. He placed "all the old instruments, as mural and mobile quadrants," to the south, in areas closest to the transepts.[5] In the other two areas he installed the large terrestrial and celestial globes, spheres and other objects related to astronomy. One-hundred-eighty maps of France, eighteen busts of the French astronomers who had worked measuring degrees on the meridian in France, in the North and on the Equator, and four busts of astronomers completed the objects on display in the room. An enormous statue of Cassini I, commissioned by Louis XVI from the sculptor Jean-Guillaume Moitte, was to be placed between the pillars at the entrance.[6]

Cassini IV wrote that he was satisfied that the restoration of the vaults had been completed in 1793, not long before he left the Observatory. However it is unlikely that the decor of the meridian room, his "museum", was finished, or even yet begun. The statue of Cassini I, commissioned in 1787, was still not completed in 1810 when his paper was published; Moitte had died in spring of that year. Some of the preparatory sketches for the sculpture must have been included among the documents and objects on display. In the museum the busts of the French astronomers who had measured the meridian arcs were also still expected to arrive. We know that on 27 July 1796 Cassini IV had the four busts of the astronomers taken to the observatory. Signs of the supports for the sculptures are still visible on the north side of the Cassini room. Were these the same busts and did he intend them for the "museum"?[7] Today,

5 Ibid., p. 54.

6 Ibid., pp. 54–55.

7 *Extraits des procès-verbaux du Bureau des Longitudes concernant l'Observatoire de Paris de 1795 à 1853*, Bibliothèque de l'Observatoire de Paris, Ms. 1022, f. 30.

the Observatoire has only two busts that seem to correspond to Cassini IV's description: Cassini III and La Condamine. It is possible that the globes made by Vincenzo Maria Coronelli (1650–1718) were placed at the spots planned for them. However, we have no evidence that instruments were installed in the Cassini room, and if so, which instruments. In any case it is unlikely that Cassini IV's plan would have been followed to the letter.

Admiral Mouchez's Museum at the Observatoire

In the 1830s the Bureau des Longitudes, although concerned with archival conservation, nevertheless lent numerous instruments belonging to the Observatoire to provincial observatories, in particular, the Toulouse Observatory. This might suggest that the Observatoire had lost interest in assembling and displaying a collection of historical instruments.

Things changed when Admiral Ernest Mouchez (1821–1892) arrived. He was the first to have great ambitions for an observatory museum. Mouchez was appointed director in 1878, after the death of Urbain Jean Joseph Le Verrier (1811–1877). He wanted to display some of the instruments still in the collections, which were not as lavish and extensive as he expected. Many instruments were out on loan and others had been lost. A magazine article published in 1883 extolled Mouchez's plans and deplored his predecessors' carelessness:

> Since the beginning of the Paris Observatory [...] many astronomical instruments have been successively used and replaced because of continual scientific progress. If we had taken care to conserve them, today we would possess an invaluable collection in the history of science.[8]

The journalist repeated a rumour, which he may have heard at the Observatoire, that many old instruments had been treated as "scrap metal". Wilhelm Eichens (1818–1884), a manufacturer of instruments, worked for many years for the observatory and must have purchased hundreds of kilograms of scrap copper.[9]

Only a small collection of beautiful instruments escaped this destruction, forgotten in one cupboard. These instruments would make up the core of the new museum's collection. Mouchez assiduously pursued the establishment of a policy to enlarge the collection. He requested photos from all the

8 "Instruments d'astronomie", *Magasin pittoresque* 53 (1885), pp. 88–90: 88 (translated by S. Keyes).

9 Ibid.

FIGURE 5.1 *The Museum of the Paris Observatoire in 1896; engraving from Alexis Lemaistre,* L'Institut de France et nos grands Établissements scientifiques: Collège de France, Muséum, Institut Pasteur, Sorbonne, Observatoire, *Paris, 1896, p. 311.*

observatories; French consular authorities were also consulted. He decided to have all of the borrowed instruments returned to the Observatoire.

Even more ambitiously, Mouchez supported the creation of a national museum of astronomy (*Fig. 5.1*), whose collections would be donated by other institutions, such as the Conservatoire des Arts et Metiers or the Bibliothèque Nationale. Obviously, the project failed, no institution being ready to give up its collections.

Establishment and Organisation of the Museum

The Mouchez museum was installed in the Perrault building. Mouchez probably pursued the idea to install Cassini's museum in the meridian hall, as we can surmise from a painting of the time (*Fig. 5.2*). It shows busts of astronomers and some remarkable instruments, for instance the Gambey meter comparator. In the west rotunda on the first floor a central display case held Habermel's instruments and many other small objects, along with the Passemant globes. Other display cases were installed in the window recesses, each one containing

FIGURE 5.2 *The hall of the museum and the meridian line in 1879; engraving from* Le Journal Illustré *16, 9 (1879), p. 68.*

objects pertaining to specific subjects investigated by the observatory in the past, for example, the decimal metric system. Portraits of nearly everyone who at one time or another had been head of the Paris Observatoire were hung in this room. The east rotunda seems to have been devoted to astronomical photography. Portraits of major scientists from foreign countries were displayed there. Instruments used by French astronomers outside of the country were also exhibited, such as Lacaille's sextant that he used in the southern hemisphere, as well as photographs of instruments and objects that Admiral Mouchez had collected from observatories all over the world. The "Grande Galerie" was also an exhibition area; it may have held very large scientific instruments. We know that John Bird's mural quadrant was returned to the Observatoire from the Toulouse Observatory in 1881. Correspondence about this restitution shows that the aim was to help develop the collections of the new museum.

Group visits were organized on the first Saturday of each month. Visitors had access to the Perrault Building to view exhibits in the Cassini room, the Grande Galerie and in the east and west rotundas. Guards accompanied groups of visitors in the gardens and watched over what were then modern instruments, including the great meridian circle, installed during the mid-nineteenth century, and the large cupola built in the 1840s at François Arago's request. The cupola housed the refractor, completed ten years later at the time of Urbain Le Verrier. Around 1885 the public could view several newly installed instruments: the sky chart, the coudé equatorial, and the large coudé equatorial, later known as the origin of the coudé focus for modern reflectors. "Messieurs les astronomes" (there were not yet any female astronomers) were asked by Admiral Mouchez, who was responsible for the instruments, to be present to explain to visitors how they functioned.

Most of the instruments and other objects that Admiral Mouchez included in the observatory museum are still in its collections. Some of the larger ones are on permanent display, such as the two mural quadrants by the famous instrument maker Bird, or Lacaille's sextant, and several of Foucault reflectors. The smaller instruments are stored according to their size.

In about 1908 the museum's floor space was reduced: the east rotunda was given to the Time Division; and in 1912, to the International Time Board (BIH). In 1926 the Great Meridian Hall was converted into an optical laboratory.

Special Events during the First Part of the Twentieth Century

Meanwhile, the museum was progressively dismantled. Instruments were removed and placed in different areas and soon forgotten. Several exhibitions and events were organised from time to time on various topics. Electricity, a rather new and contemporary scientific discovery, was for a while a favourite topic as it had very early come into general use at the Observatoire.

In 1935, on the occasion of the General Assembly of the International Astronomical Union in Paris, a garden-party was organised on the grounds of the observatory. The participants could view the historical instruments on display in the buildings, as well as the instruments in use. At that time a half dozen instruments were on the grounds and two were in the great domes on the roof.

In 1946 a colloquium was organised at the Observatoire, whose director was then André Danjon (1890–1967). The year 1946 was the centenary of the discovery of Neptune. Le Verrier had predicted its position, and a special session was held in the large amphitheatre of the Sorbonne. At the same time, the Observatoire held an exhibition of instruments, objects, and manuscripts in the Grande Galerie on the first floor (*Fig. 5.3*).

FIGURE 5.3 *The exhibition,* Le Verrier et son temps, *in 1946.* (COLLECTIONS OF THE BIBLIOTHÈQUE DE L'OBSERVATOIRE DE PARIS).

During this period the monthly visits still took place, but were organised differently. Twelve young astronomers (from the beginning of the twentieth century there were both men and women) each chose a month to take visitors around and to explain the instruments displayed on the ground floor (the speaking clock and Foucault's two large telescopes), as well as those on the first floor and on the grounds. The older astronomers were asked to give tours that corresponded to specific requests.

Groups of about thirty visitors were accompanied by guards wearing dark blue uniforms, labelled "OP" in gold. The guards unlocked doors and made sure that no one manipulated the instruments. Most of the guards were very fond of the instruments and were allowed to move the larger ones under the supervision of the astronomers.

The year 1967 was the tercentenary of the Observatoire. An exhibition was organised by the new curator, who was also head of the library, which included manuscripts and instruments in its collections. Several astronomers also lent their expertise. An exhibition catalogue was published. The minister in charge of the Observatoire opened the exhibit on 21 June.

The Museum: 1970 to 2010

At the end of the 1960s, the museum in the west rotunda was also dismantled. The rooms were turned into offices. In 1976 a number of small pieces, mostly sundials, were stolen from the display cases that had been bought for the tercentenary. For several months the display cases remained empty, until Pierre Charvin was appointed head of the Observatoire in 1981. One of the historians at the observatory, in cooperation with the conservator of scientific events, was asked to reorganise the visits to include a "foreign policy" aspect, thus to foster exchanges with researchers from other countries. Some instruments were moved and placed elsewhere in order to modify the route of the guided tours. Every year, the curator determined the theme of the display in relation with some event or discovery of significance to the Observatoire, or related to a symposium organised by the institution. Nevertheless, the museum was at this time limited to the Grande Galerie Hall. Its floor-space was enlarged by the Picard Hall that earlier had housed the Observatory Council. Only copies of documents were shown in the display cases for security reasons.

Although matters did not look bright for the museum, some exhibits were held in the Perrault building. A colloquium was organised in 1982 on the occasion of the tercentenary of the death of Jean Picard (1620–1682), considered to be the founder of astrometry and geodetic astronomy in France. At that time, instruments, documents and manuscripts were displayed in the Grande Galerie and in the "Picard Room".

In 1983, technology offered another chance to the museum. The optical laboratory seemed to be too small to house the new mirrors required by researchers in the field of astrophysics; the laboratory was moved outside the observatory. The Great Meridian Hall of Cassini I was completely emptied and restored. A larger museum was discussed, but it was considered that a better use would be to organise large exhibitions there. This would be in addition to the reorganisation of the permanent exhibit on the first level, which was visited regularly. The large room on the second floor was called the "Salle Cassini".

The 'Golden Age' of Temporary Exhibitions

The first exhibition organised in the Salle Cassini occurred in 1983 to celebrate the new realisation of the length of the metre in which the Observatoire had played a role. Called *Longueur et Temps: De la vitesse de la lumière à la définition du mètre*, it was an impressive exhibit that included documents, manuscripts, and delicate or fragile instruments presented in the display cases purchased in

1967. An exhibition catalogue was published in cooperation with the library of the observatory.

The same year marked the fiftieth anniversary of the "*horloge parlante*". An Open House on Saturday and Sunday featured such a clock. Instruments related to time were exhibited in the great hall on the ground floor, in what is now called the Esclangon Room. The three talking clocks available at the time were displayed in the room. The 1933 clock no longer worked, but a voice recording on an audio cassette was used. There were two 1965 clocks and one from 1975. These three clocks were in working order, and one could be reached by telephoning 3699. The beep was provided by the Time Service. There were so many visitors that the queue stretched all the way to the avenue Denfert-Rochereau. To distract visitors during their long wait, three astronomers, standing near the entrance to the great hall, told the story of the speaking clock and of the observatory.

In 1985, the UK and the Netherlands asked the Observatoire to organise the International Astronomical Union Symposium for 1987. The meeting was called "Mapping the Sky". Those responsible at the observatory decided to organise a public exhibition to coincide with the Symposium. The title of the exhibition was *La Mesure du Ciel: De la plaque photographique aux techniques spatiales* (*Fig. 5.4*). Large panels for display purposes were installed in the Cassini room; a selection of objects, instruments, documents, and manuscripts was presented. An office that was supposed to duplicate Le Verrier's was installed. The desk and portfolio were Alexander von Humboldt's, the armchair had belonged to Arago. An old ink-bottle and an old pen were loaned by an elderly astronomer. A colleague brought a petrol lamp from his apartment. Many leaves photocopied from Le Verrier's manuscripts were placed on the desk, while others were exhibited in a dust-bin, as if Le Verrier himself had tossed them out. One day, the person in charge of keeping an eye on the office arrived and found the dust-bin empty; the cleaning woman had done her job so well that new photocopies had to be made for the display. The exhibition, as well as the catalogue, was the result of cooperation among the scientists, the conservator, and the librarians. The Centre National d'Études Spatiales (CNES) provided a model of Ariane.

A smaller exhibition, *Newton 87*, was arranged by the library in the Picard Room, following a suggestion by René Taton (1915–2004), the noted French historian of science. The year 1987 was the 300th anniversary of the first edition of Newton's *Principia*. The exhibition was chiefly books and papers; the instruments on display were in the Grande Galerie.

Except for a short interruption, exhibitions were organised nearly every year from 2001 until 2006. To celebrate the beginning of the twenty-first century

FIGURE 5.4 *Presentation of the exhibit,* La Mesure du ciel, *in 1987.*

and the third millennium, an exhibition *Temps mesurés-Temps démesurés* was presented in 2001. An exhibition dedicated to Léon Foucault, *Léon Foucault, le miroir et le pendule*, was held in 2002, and another was dedicated to Arago in 2003 (*Pl. 6*). In 2004 an exhibition entitled *Au-delà de la Terre* presented solar system exploration through the ages. The World Physics Year in 2005 provided the occasion to present the exhibition *"c" à Paris* about early experiments on the velocity of light (Foucault, Arago, Fizeau, Cornu) and to shoot lasers between the Observatoire and Sacré-Coeur (*Fig. 5.5*).[10]

Historical Exhibitions and AMA 2009

Several events led to an interruption of the exhibitions held in the Cassini room, which had always been open to the public. Efforts to promote the observatory's heritage had paid off and funding was made available for the preservation of collections. Starting in 2006, an intense period of cataloguing, restoration and conservation of the patrimonial collections commenced. These endeavours, however, were incompatible with large shows and with exhibitions that would involve finding volunteers to greet visitors. Hence exhibitions geared to a more restricted public (students, groups, visitors to the observatory) and linked exclusively to historical collections are held in the Grande Galerie each year. These exhibitions began in 2004 with one entitled *Méchain et la longueur du metre*. In 2006 they became regular events with an exhibition on the evolution of astronomical instruments traced through the permanent collection. In 2007 an exhibition was dedicated to two astronomers, Jérôme Lalande (1732–1807) and Maurice Loewy (1833–1907), on the occasion of the bicentenary and centenary of their deaths, respectively. The "BIPM Metrology Summer School 2008" a year later was dedicated to the metric system. In autumn of the same year, an exhibition on the astrometric satellite Hipparcos was displayed. It highlighted the space project's archives that the European Space Agency (ESA) had given to the observatory. During the World Astronomy Year in 2009, the Hipparcos exhibition remained on view until it was decided to remodel the Picard Room to enlarge the gallery space. It has now been completed and the room reopened in September 2010.

This great diversity of temporary shows open to the general public enables the permanent exhibition to be renewed each year in the Grande Galerie, with large exhibitions in the Cassini room, including external loans, on subjects specific to the Observatoire. These two kinds of exhibitions are complementary. The latter facilitate more contact with scientists thus ensuring supervision of the events. However, these exhibitions are very complicated to organise given the current state of affairs, due to the difficulty of organising a group of volunteers to deal with visitors in an institution where research is the core activity.

10 Some of the exhibitions may still be viewed as virtual at http://expositions.obspm.fr/ (accessed 17 March 2014).

FIGURE 5.5 *Poster for the exhibition,* "c" à Paris, *in 2005.*

FIGURE 5.6 *The exhibition,* L'Astronome du roi et le satellite, *in 2012.*

It should be mentioned here that the Observatoire has been closely associated with the "Journées du patrimoine", or Heritage Days, since their creation in 1984, on the initiative of Jack Lang, then the Minister of Culture. They were called "Journées portes ouvertes dans les monuments historiques" (Open house: Historical monuments), and took place on the third Sunday in September. In 1991 the Council of Europe put them on a European level. They have always been very successful. Several thousand people visit the Observatoire on those two days (up to 8,000 in 2010!) to view the instruments and the exhibitions.

Since the arrangement of a new room, so called "Picard Room", in 2010–2011, exhibitions are regularly held every year. The astronomers Urbain Le Verrier in 2011, Gian Domenico Cassini in 2012 (*Fig. 5.6*), La Caille in 2013 were honored in this way. These exhibitions attract a varied and relatively important public: more than 10,000 visitors a year, as well scholars, amateur astronomers, scientists, teachers, and of course the general public.

Conclusion

Today, as in Cassini IV's time, it is not always easy to reconcile the ongoing scientific activities with the historic and patrimonial missions of the Observatoire, mainly for financial reasons. On the other hand, public exhibitions provide an opportunity to put into perspective the past and present quests and to show old and contemporary documents side by side, the science of the past and the projects of future. It is the mark of an exceptional continuity (350 years of science announces the title of a book newly published about this long and extraordinary history!)[11] and it indicates the true identity of the "Observatoire de Paris".

---

11 Laurence Bobis, James Lequeux (eds.), *L'Observatoire de Paris: 350 ans de sciences*, Paris, 2012.

CHAPTER 6

Looking at Scientific Instruments on Display at the United States Centennial Exhibition of 1876

*Richard L. Kremer**

Introduction

The nineteenth-century phenomenon of the 'universal exposition' has been thoroughly studied by cultural historians. Beginning with London's Great Exhibition of the Works of Industry of all Nations held at the Crystal Palace in 1851, these mega-events provided opportunities for industrial nations to represent themselves (and their empires) to each other and to define themselves in opposition to 'exotic' non-industrialized societies; for enthusiasts of technological, scientific and cultural progress to valorize their narratives and dreams; for private companies to advertise their products and 'do business' of buying and selling; and for hundreds of thousands of visitors to gawk at the world's industrial merchandise on display in massive exhibition spaces. The industrial and imperial world, barely a century old, found in the universal exposition a new medium for self-representation.[1]

* I would like to thank participants at the *XXIX Symposium of the Scientific Instrument Commission* in Florence for their questions and suggestions, and especially Paolo Brenni, David Pantalony, and Marc Rothenberg for their help and comments on earlier drafts of this essay.

1 For some of the many studies, see Robert W. Rydell, *All the World's a Fair: Visions of Empire at American International Expositions, 1876–1916*, Chicago, 1984; Paul Greenhalgh, *Ephemeral Vistas: The Expositions Universelles, Great Exhibitions, and World's Fairs, 1851–1939*, Manchester, 1988; Brigitte Schroeder-Gudehus, Anne Rasmussen, *Les fastes du progrès: Le guide des expositions universelles, 1851–1992*, Paris, 1992; Anne-Laurie Carré et al. (eds.), *Les expositions universelles en France au XIX*e *siècle: Techniques publics patrimoines*, Paris, 2012; Robert W. Rydell, Nancy E. Gwinn (eds.), *Fair Representations: World's Fairs and the Modern World*, Amsterdam, 1994; John E. Findling, Kimberly D. Pelle, *Encyclopedia of World's Fairs and Expositions*, Jefferson (North Carolina), 2008; Winfried Kretschmer, *Geschichte der Weltaustellungen*, Frankfurt, 1999; Linda Aimone, Carlo Maria Olmo, *Le esposizioni universali, 1851–1900: Il progresso in scena*, Turin, 1990; Paolo Colombo, *Le esposizioni universali: I mestieri d'arte sulla scene del mondo, 1851–2010*, Venice, 2012. Although in France, Spain, the Low Countries and the United States (after 1876) these events were called 'expositions' but in Britain and its

 | DOI 10.1163/9789004264403_007

As Jim Bennett, Paolo Brenni and others have noted, scientific instruments already in 1851 took their place amidst the myriad of products on display at these expositions.[2] In the early expositions, just decades after the industrial production of scientific apparatus had begun, first in England, then in France and German-speaking regions, instrument makers and their international agents sought to display their latest and best wares and, of course, to win medals of excellence awarded by the judges. Not until 1876, at the first universal exposition held in America, would United States' instrument makers prominently feature in these displays.[3] On the one hand, the inclusion of scientific instruments signaled nothing new for the culture of the expositions; displays of instruments contributed to breathless narratives of progress and to international competitiveness just as did displays of steam engines, sewing machines or high-art paintings. On the other hand, the arrival of scientific instruments at the expositions did reflect a new awareness, among the instrument-makers, of the importance of international markets for their industry.

In this essay, my goal is not simply to describe the scientific instruments on display at the United States Centennial Exhibition, held in Philadelphia from May through November of 1876. Rather, I want to consider how various groups 'saw' those instruments. I want to ask how catalogue readers, the exhibition's judges, international writers of the exposition's 'panoramic literature', and the general public described the scientific instruments they saw on display. Ten million people visited the Philadelphia exposition (one-fifth of the entire pop-

colonies were denoted 'exhibitions', the two terms apparently carried (and carry) the same cultural meaning.

2 Jim Bennett, *Science at the Great Exhibition*, Cambridge, 1983; Paolo Brenni, "19th Century Scientific Instrument Advertising", *Nuncius* 17 (2002), pp. 497–513; Brenni, "La science française au Crystal Palace", in Patrice Bret (ed.), *Les techniques et la technologie entre France et Grande-Bretagne, XVIIe–XIXe siècles*, Paris, 2010, pp. 255–265; Brenni, "Effects spéciaux! Science et technique au service des attractions de l'exposition universelle de 1900", in Carré et al. (eds.), op. cit. (n. 1), pp. 75–90.

3 An exhibition held in 1853 in New York was small (13 nations, 700,000 total visitors), privately funded (it failed financially), and often is not considered among the 'universal expositions' of the nineteenth century. See Greenhalgh, op. cit. (n. 1), p. 38. Its 31 classes of exhibited articles did include one for "philosophical instruments, and products resulting from their use (e.g., daguerreotypes, etc.), maps and charts, horology, surgical instruments and appliances," but most of those exhibitors were inventors and entrepreneurs rather than makers of traditional philosophical or mathematical instruments. Cf. *Official Catalogue of the New York Exhibition of the Industry of All Nations*, New York, 1853, p. 19; Robert C. Post, "Reflections of American Science and Technology at the New York Crystal Palace Exhibition of 1853", *Journal of American Studies* 17 (1983), pp. 337–356; Ivan D. Steen, "America's First World's Fair", *New York Historical Society Quarterly* 47 (July 1963), pp. 257–287.

ulation of the United States in 1876). Never before had so many people seen scientific instruments in such a concentrated fashion; on one day, 28 September 1876, over 275,000 people streamed through the exposition. What did they see? And what do their reports tell us about the cultural meaning of science and its instruments in the mid 1870s?

The Catalogue Readers

"I am the Catalogue of the Great Exhibition. You are the public. I intend to have some private talk with you, and pour into your ear the story of my early life". So opened a short story appearing in an 1851 issue of *Household Words*, the London literary journal edited by Charles Dickens. Authored by Henry Morley, a frequent writer in Dickens' journal, "A Catalogue's Account of Itself" explained the work required to categorize the world as displayed in the Crystal Palace Exhibition. According to the story, all potential exhibitors were asked to complete a printed form, describing their product. Six compilers at the publisher then worked five months to impose order, clarify language, and combine 15,000 entries into a single catalogue of only 323 pages.[4] It was the catalogue-makers, Dickens and Morley realized, who brought order to the international exposition; visitors could experience an exposition only after a catalogue had generated its categories and frameworks of meaning.

The 'official' catalogue divided the world of the Philadelphia Exposition into 7 departments,[5] 36 groups, and 739 classes. Two groups, both located in Department III (education and science), included instruments: Group 24 (which we shall here ignore) encompassed "instruments and apparatus of hygiene, medicine, surgery, prosthesis"; Group 25 presented "instruments of precision, research, experiment, and illustration, including telegraphy and music". Eight classes further subdivided Group 25:

4 Henry Morley, "The Catalogue's Account of Itself", *Household Words* 3 (1851), pp. 519–523.

5 Mining and metallurgy (2100 exhibitors), manufactures (8800), education and science (2500), art (4900), machinery (2300), agriculture (10,200), and horticulture (100). See J.S. Ingram, *The Centennial Exposition Described and Illustrated*, Philadelphia, 1876, p. 641; United States Centennial Commission (USCC), *Official Catalogue, Complete in One Volume*, Philadelphia, 1876. Slightly different wordings appear in the official post-exposition publication that included judges' citations, visitor numbers, and much more. Cf. USCC, *International Exhibition 1876*, Washington (D.C.), 1880–84, 9 vols. For a Foucauldian study of the ordering of the Philadelphia exhibition and the initially proposed classification scheme with ten departments, see Bruno Giberti, *Designing the Centennial: A History of the 1876 Intenational Exhibition in Philadelphia*, Lexington, 2002.

Class 320: Instruments of precision, and apparatus of physical research, experiment, and illustration;
Class 321: Indicating and registering apparatus, other than meteorological; mechanical calculation;
Class 322: Weights, measures, weighing and meteorological apparatus;
Class 323: Chronometric apparatus, including astronomical clocks, church clocks, ordinary commercial clocks, watches, chronographs, metronomes;
Class 324: Optical and thermotic [sic] instruments and apparatus;
Class 325: Electrical apparatus;
Class 326: Telegraphic instruments and methods;
Class 327: Musical instruments and acoustic apparatus.

The official catalogue, printed before the exposition opened,[6] lists a total of 639 separate instrument makers or firms, including 249 musical, exhibiting in Classes 320–327 (*Fig. 6.1*). An attentive catalogue reader might, first, have been astonished to see the global reach of instrument making by 1876. Nearly every nation exhibiting in Philadelphia had local instrument-making industries. London and Paris, long the leading centers of scientific instrument production, now faced considerable competition. Second, our catalogue reader would surely have noticed the linking of scientific instruments and musical instruments (a common pattern at the world expositions), and the placement of both in the department of education rather than manufacturing. Scientific instruments, to the Philadelphia catalogue-writers, belonged to the world of education and science, not the commercial spaces of the factory and the telegraph office or the practical spaces of the surveyor, shipbuilder, or bookkeeper. Third, the catalogue reader would have detected the large percentage of musical makers among the exhibitors from the USA, Germany, Britain and France. In each of these nations, the structure of the scientific as well as the musical instruments industry—many smaller rather than a few monopolistic firms—is striking.

Our catalogue reader undoubtedly would not have been surprised to see many leading firms represented at the Philadelphia exposition. The American exhibitors included E.S. Ritchie & Sons, Keuffel & Esser, W. & L.E. Gurley, Heller & Brightly, Wm. J. Young & Sons, James W. Queen & Co., Joseph Zentmeyer, Bausch & Lomb Optical, Western Electric, Thomas A. Edison and Western Union Telegraph. French makers included Deleuil, Bréguet & Co., J. Duboscq,

6 In other words, not all of these announced exhibitors may have actually journeyed to Philadelphia.

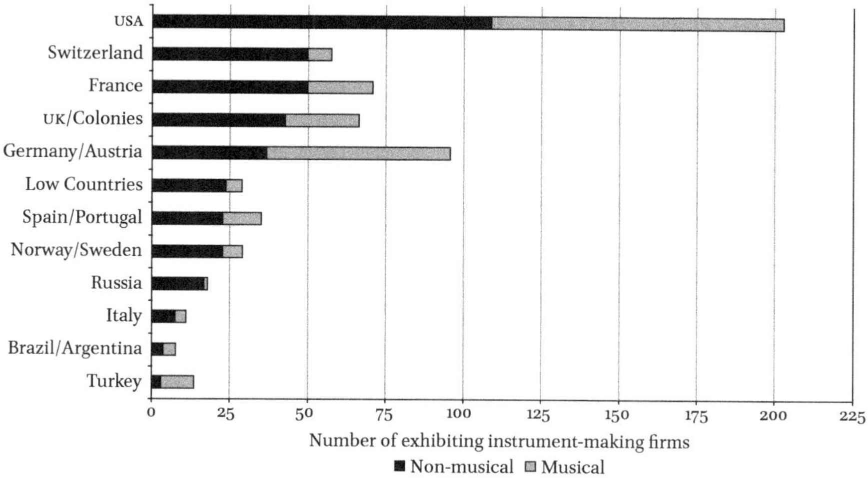

FIGURE 6.1 *Number of makers exhibiting in Philadelphia from Classes 320-327, by nation. United States Centennial Commission,* Official Catalogue, *Philadelphia, 1876, pp. 329–373.*

A. Nachet, and Ernst Panier; London makers included Negretti & Zambra, R. & J. Beck, Henry Couch, Siemens Bros., and James Joseph Hicks. Only a few German makers appear, however: G.A. Schultze, F. Sartorius, Voigtlaender & Sohn. As we shall see below, many observers wondered why leading German makers, such as Schmidt & Haensch, Carl Zeiss, C.A. Steinheil, Siemens & Halske, E. Leybolds Nachfolger, or R. Fuess, were not represented.

And our catalogue reader, if able to separate the wheat from the chaff, might well have been intrigued (or bemused) by some of the lesser-known makers who claimed to be offering new inventions or highly specialized apparatus. William A. Rogers, assistant at the Harvard College Observatory, for example, announced an "automatic machine for ruling microscopic lines on glass or metals" to make diffraction gratings; H.A. Clum of Elmira, New York, had an "aëllograph or storm writer for pre-determining storms"; George Mehr of Philadelphia had "patent sun dials made of marble" for sale; J.W. Hill of Waterville, Kansas boasted of a "clock that will run 100 years without winding [...] and be perfectly accurate";[7] F.W. Meyer of Montreal had an "apparatus for determining inaccurate points" (I have no idea what this might have been); Paul Emile Le Boulengé of Liége presented a new "telemeter for the battle-field"; P.J. in de Betou of Delft an "instrument measuring 100th part of a millimetre";

7 Ingram, op. cit. (n. 5), p. 303, described the clock as "the nearest to perpetual motion that we believe has ever yet been attained".

Martin Wiberg of Stockholm a "bull-dog apparatus for deep-sea soundings"; Carlo Ponti of Venice a "camaziale and megalithoscope (optical instruments)"; and George Diakoft of St. Petersburg offered an "arithmometer, new system". The large, well-established firms exhibited extensive collections of products, described in the official catalogue simply as "scientific instruments"; the small and the idiosyncratic makers used that catalogue to advertise exotic and novel apparatus. Our catalogue reader undoubtedly marveled at the breath-taking diversity of material lumped together under Group 25, even before setting foot in the Main Exhibition Building, a massive space 1,880 feet long, 464 feet side, and 70 feet high, to take in the scientific instruments on display.

The Panoramic Writers

Beyond the official catalogues, several other distinctive literary genres emerged to celebrate or otherwise respond to the culture of the nineteenth-century international exposition.[8] One genre, denoted as 'panoramic' by Walter Benjamin and deriving from earlier romantic travel narratives, sought to take in the whole spectacle, to create a visual and verbal diorama in a medium that was not exactly journalistic, literary, pictorial, or scientific, but somehow combined all these approaches. One literary critic ridiculed these writers as "journalistic Knights of Industry with their eternal lead pencils, these encyclopedic reporters of everything, these Universal Geniuses, who with equal inspiration, equal knowledge and equal naiveté judge machines and cigarette holders, [Peter von] Cornelius and rubber, [Paul] Delaroche and leather boots, lighthouses and matchsticks, thumbtacks and patent soles, Lyon silk and rape oil".[9] Of course the Philadelphia Exhibition attracted its share of panoramic writers; but did they see any scientific instruments, such small, quiet and drab artifacts when placed against the massive steam engines in Machinery Hall, the 'foreign' exhibitions from Asia, Africa and Russia, or the exotic plants in the Horticultural Hall?

As we might expect, the exposition filled American panoramic writers with "a thrill of patriotic pride", as William Dean Howells wrote in the July 1876 issue of the *Atlantic Monthly*. Writing mostly about the high art and paintings on

8 For a useful overview of these genres, see Robert Brain, *Going to the Fair: Readings in the Culture of Nineteenth-Century Exhibitions*, Cambridge, 1993, pp. 9–14.

9 Quoted ibid., p. 12. Cf. Walter Benjamin, *The Arcades Project* (translated by Howard Eiland and Kevin McLaughlin), Cambridge, 1999, p. 531.

display, Howells also meditated eloquently on the giant Corliss engine prominently performing at the center of Machinery Hall, describing:

> an athlete of steel and iron with not a superfluous ounce of metal on it; the mighty walking-beams plunge their pistons downward, the enormous fly-wheel revolves with a hoarded power that makes all tremble, the hundred of life-like details do their office with unerring intelligence [...]. All that Great Britain and Germany have sent is insignificant in amount when compared with our own contributions; the superior elegance, aptness, and ingenuity of our machinery is observable at a glance.[10]

It was American machinery that impressed Howells; indeed, the Corliss engine was undoubtedly the centerpiece of the exhibition for most visitors (*Fig. 6.2*). Howells did not mention scientific instruments in his essay.[11]

Other American panoramic writers did notice the instruments. J.S. Ingram, who wrote for a Philadelphia publisher and otherwise is quite unknown in late nineteenth-century American life, was impressed by the self-registering meteorological instruments shown in the U.S. Government Building. Developed by staff at the U.S. Signal Service Corps (forerunner of the federal weather service), these thermometers, anemometers, barometers, etc., would "make a record fuller and more accurate than any made by human observers". These instruments were "thought to be the most reliable of any invented, and are in every respect the most simple in construction".[12] For Ingram, simplicity and reliability characterized the American approach to instrumentation.

In the displays of Group 25, only telegraphy attracted Ingram's attention. Printing telegraphs, sounders for telegraph lines, and burglar alarms all illustrated the "great utility of electricity in its practical application to the domestic purposes of life, and to the wants and requirements of public companies, hotels, mercantile and manufacturing establishments".[13] The scientific instruments of Group 25 do not appear in Ingram's panoramic review; neither do

10 William Dean Howells, "A Sennight of the Centennial", *Atlantic Monthly* 38, 225 (1876), pp. 92–107. Cf. David Nye, *American Technological Sublime*, Cambridge, 1994, pp. 120–123.

11 For another panoramic account emphasizing fine arts and domestic products and ignoring science and its instruments, see "Characteristics of the International Fair", *Atlantic Monthly* 38 (1876), pp. 85–91, 233–239, 350–359, 492–501 and 732–740; *Atlantic Monthly* 39 (1877), pp. 94–100 and 284.

12 Ingram, op. cit. (n. 5), pp. 122, 124. For the nineteenth-century quest for 'mechanical objectivity', see Lorraine Daston, Peter Galison, *Objectivity*, New York, 2007.

13 Ingram, op. cit. (n. 5), p. 291.

FIGURE 6.2 *Corliss engine, Machinery Hall, by George H. Corliss, Providence, Rhode Island* (PRINT AND PICTURE COLLECTION, FREE LIBRARY OF PHILADELPHIA, CEDC 042815. COURTESY OF FLP AND BRIDGEMAN ART LIBRARY).

the telephones exhibited in Philadelphia by Alexander Graham Bell and Elisha Gray.

The American panoramic writer James D. McCabe, known for his popular books on the U.S. Civil War and American history, saw more instruments than did either Ingram or Howells. As he wandered through the exhibitions, he noticed not only the telegraphic apparatus but also the "interesting collection of scientific instruments" displayed by the Stephens Institute of Technology (a school in New Jersey), an equatorial transit made by Fauth & Co. of Washington (D.C.), and the excellent "collection of scientific and philosophical instruments" in the French section, "representing the best work of the best makers". Belgium's scientific apparatus were also "excellent", but those from Austro-Hungary only "credible". The Swiss watches drew McCabe's praise, as did their "excellent and extensive" display of mathematical, scientific and philosophical instruments". The Russian "exhibit of mathematical and philosophical instruments was small but very interesting". Despite this internationalism, McCabe concluded his massive tome with the ritualistic, chauvinistic praise of American accomplishments. "In the eyes of the world we [Americans] have attained a rank never according us before [...] the astonishment and wonder of our visitors from abroad at our resources and accomplishments was great and universal".[14]

Thus, the scientific instruments on display in Philadelphia registered with the panoramic writers I have sampled but clearly attracted far less attention than did the cornucopia of industrial and agricultural products and the fine and domestic arts that filled the exhibition buildings.

The National Commissioners

As part of the international competition undergirding international expositions, national governments and trade or industrial organizations regularly deputized commissioners to write reports on the competitive status of their nation in the world's economy. As is well known, Britons worried greatly, after the 1851 Crystal Palace Exposition, that their nation was losing the industrial superiority it thought it had held since the eighteenth century. The Philadelphia exposition similarly attracted delegations of worried commissioners from abroad, with British observers again wringing their hands:

> The competition at Philadelphia was not altogether satisfactory to us. It is true that every nation has an advantage in exhibitions held within its own area; but the products of the industry of the U.S. surpassed our own oftener than can be explained by circumstance. It appeared as if there was a greater economy of labour habitually practiced in the States [...] [and] there was evidence of the more constant presence of a presiding mind superintending every process of industry [...] they revealed the application of more brains than we always have at our command.[15]

What did these national commissioners write about the scientific instruments on display?

Perhaps the most widely discussed national comparisons came from Franz Reuleaux, a well-known mechanical engineer, director of the Gewerbe-Akademie, a technical school in Berlin, and chair of the German contingent

14 James D. McCabe, *The Illustrated History of the Centennial Exhibition*, Philadelphia, 1876, pp. 379, 408, 418–26, 463 and 893.

15 *London Times*, 1 March 1877.

of judges at the Philadelphia Exposition.[16] His *Briefe aus Philadelphia*, appearing in German (and London) newspapers during the summer of 1876, sharply criticized German manufactures as "*billig und schlecht*" ("cheap and nasty" as translated by the *London Times*). He blasted German exhibitors for showing only Bismarck memorabilia and Krupp cannons, artworks that lack taste and poorly made industrial products (especially machine tools and steam engines, when compared to U.S.-made versions). He criticized German industry more generally for seeking to compete in international markets by lowering price rather than raising quality. And he noted, perceptively, that perhaps the high American tariffs (averaging 41% *ad valorem* in 1875)[17] kept many European exhibitors away. Reuleaux's letters prompted a national debate in the newly unified Germany about its political economy; but he wrote nothing, as far as I can see, about scientific instruments *per se*.[18]

An Austrian judge viewed American industry less favorably. In every branch of the Exposition one saw the "predominantly practical direction of the Americans and their industry, with theory and its accessories ignored or poorly handled. This was especially apparent in precision mechanics, whose products support the sciences as well as the needs of daily life".[19]

The French commissioners reported specifically on precision-mechanics, giving the most extensive commentary on the displayed scientific instruments that I have found (they also traveled around the United States and reported broadly on American economic, social, educational and political life). Not surprisingly, they described the instruments as they were laid out in the exhibition halls, *viz.*, nation by nation. In the American section, they praised the mechanical models by Keuffel & Esser, Alvan Clark's large equatorial telescopes, and the variety of surveying and geodetic apparatus offered by Gurley,

16 Reuleaux had served as a judge representing Prussia at earlier international expositions in London (1862 and 1867), Vienna (1873) and Dublin (1873). *Neue Deutsche Biographie*, Berlin, 1953–, s.v., "Reuleaux, Franz".

17 R.C. Post, *1876: A Centennial Exhibition*, Washington (D.C.), 1976, pp. 179–180.

18 Franz Reuleaux, *Briefe aus Philadelphia*, Braunschweig, 1877. For responses to Reuleaux, that also do not discuss scientific instruments, see H. Heine, *Professor Reuleaux und die deutsche Industrie: Eine Skizze auf Grundlage amerikanischer sowie deutscher Beobachtungen und Erfahrungen*, Berlin, 1876; Georg Hirth (ed.), *F. Reuleaux und die deutsche Industrie auf der Weltausstellung zu Philadelphia*, Leipzig, 1876; Friedrich Goldschmidt, *Die Weltausstellung in Philadelphia und die deutsche Industrie: Drei Vorträge*, Berlin, 1877.

19 Johann Spácil, *Wagen, Dynamometer und Materialprüfungs-Maschinen, Bericht über die Weltausstellung in Philadelphia 1876, Heft 19*, Vienna, 1877, p. 7.

Wm. J. Young & Sons, and Heller & Brightly. In general, they did not find "superior quality" among the American mathematical instruments, but thought that the United States could supply itself in this field. They urged French makers not to abandon large telescope making to the Alvan Clark (the French astronomer Urbain Jean Joseph Le Verrier had just written Clark asking if he could make a 1.2 meter mirror for the Paris Observatory); concerning geodetic instruments they worried that "we must fight without delay if we want to maintain our reputation and our exports".[20]

On the other hand, E.S. Ritchie, "one of the most important houses in the USA" whose workshops they had visited in Boston and Brooklyn, was less able to compete against European imports in the broad market of pedagogical apparatus for schools and universities. With labor costs one-third higher than in Europe, Ritchie was forced to simplify and to replace brass with wood wherever possible. This makes Ritchie apparatus less durable; but they are "very neat" and adequate for classroom demonstrations. Elsewhere the French observers labeled as the "American type" such "cheap but neat" scientific instruments.[21]

The French commissioners praised London optical makers R.J. Beck, J.H. Dallmeyer, Negretti and Zambra, and Andrew Ross for their fancy microscopes with multitudes of accessories, worrying that French microscopes appear rather dull and modest in comparison (previously, many French observers had judged foreign instruments to lack the elegance of the Parisian-made versions). Germany was poorly represented, as many of its leading precision-makers had elected not to exhibit in Philadelphia. The precision-balance maker, Sartorius of Göttingen, did exhibit but to the French eyes, its aluminum balances were "unsightly" and "too complicated". Japanese instruments they judged to be poorly made and technically lacking, as were the Russian instruments. Portugal drew praise for its Industrial Institute in Lisbon, founded in 1852 to train scientific instrument makers. Canada too was praised for Ontario's province-wide efforts to provide instruments, both Canadian and foreign-made, for elementary and secondary school classrooms.[22]

Of course the French commissioners reported extensively on the French exhibits in Philadelphia. Deleuil's balances were the "best displayed [...]

---

20 F. Maquaire, Th. Ponthus, A. Harlé, *Exposition universelle de Philadelphia, 1876: Rapport des délégués mécaniciens en précision, présenté à M. le Ministre de l'agriculture et du commerce*, Paris, 1879, pp. 38–80.

21 Ibid., pp. 154–159.

22 Ibid., pp. 81–85, 93–100 and 169–174.

[their] precision is perfect". Rudolph Koenig's massive exhibit of acoustic apparatus "was one of the most interesting in physics, especially for the variety of apparatus". Bardou et Fils offered several significant telescopes. Yet many of the French instruments seemed to lack the "cachet and brilliance" of similar British- or American-made products, a deficiency attributable perhaps, the commissioners speculated, to damage during transport or the poor arrangement of vitrines in the French sections. The French commissioners concluded, more fundamentally, that U.S. tariffs and special societies, institutes and public lectures for training technical workers were giving advantages to American and British instrument-makers. Only with better education for its workers, they suggested, could French makers compete in the global market.[23] This rhetoric of French makers falling behind would grow for the rest of the century.

The Judges

If foreign commentators detected an 'American type' of scientific instruments, they also grumbled about the uniquely 'American System of Awards' they found in Philadelphia. Rather than granting gold/silver/bronze medals or other ranked honors, the organizers deliberately crafted an 'American system' of citations in which judges would discuss noteworthy exhibits (in signed comments) but not otherwise rank them by quality. The judges issued a total of 385 individual citations in Group 25, a total representing more than 60 percent of all exhibitors in that group.[24] First used in Philadelphia, this supposedly more egalitarian system of distributing many citations but no gold metals satisfied few; it would not be continued at subsequent international expositions of the nineteenth century.[25]

Fifteen judges evaluated the exhibits of Group 25. They elected as their chair William Thomson, professor of natural philosophy at the University of Glasgow and by 1876 undoubtedly the world's best-known physicist; in 1866, Thomson had been knighted for engineering the first successful Atlantic submarine telegraph cable. The U.S. judges (excluding those for the musical instruments) included Joseph Henry, founding Secretary of the Smithsonian Institution; Frederick A.P. Barnard, astronomy enthusiast and President of Columbia University; Julius E. Hilgard, superintendent of the U.S. Coastal Survey and in 1875 president of the American Association for the Advancement of Science;

23 Ibid., pp. 174–178 and 321–356.

24 USCC, *International Exhibition*, cit. (n. 5), v. 7, pp. 451–517.

25 Giberti, op. cit. (n. 5), pp. 154–174.

and James Craig Watson, director of the Detroit Astronomical Observatory owned by the University of Michigan. Foreign judges (also excluding music) included a professor of economics, geography and history at the Collège de France; an Austrian railway engineer; a Swiss watchmaker; and Professor Reuleaux (see above) from Berlin. These judges all had extensive hands-on experience with scientific instruments, mostly in practical rather than theoretical contexts. What did they see in the instruments displayed in Philadelphia?

As was typical at the international expositions, the judges examined the exhibited instruments not only visually but also in some cases through hands-on empirical trials. Among other items, the chronometers and watches, musical instruments, balances, photographic lenses, and water meters were put through their paces. Among these trials was Alexander Graham Bell's legendary demonstration of his duplex telegraph and telephone to judge Thomson and the Brazilian emperor, Dom Pedro.[26]

The judges issued their reviews in two formats, general overviews of each class and individual citations for specific makers. Several themes appear in the former. Most notably, the judges repeatedly decried the lack of novelty in the exhibited instruments. The showcases might be filled with many types of surveying and geodesic apparatus, microscopes, projection lanterns, self-registering meteorological instruments or marine chronometers and watches, but very few items drew praise for innovative designs or materials. Only a barometer by Dmitri Mendeleev of St. Petersburg (author of the periodic table in 1869), designed to measure slight changes in elevation, a spectrometer by Duboscq of Paris, and a short-arm balance by Sartorius of Göttingen, designed for speed of use rather than precision, seemed groundbreaking to the judges. Second, judge Henry emphasized the practical applications in the electrical class. Indeed, he mentioned only one research apparatus (galvanometers that, however, display "no new principles") but described with great flourish the myriad of dynamos, batteries, motors, submarine cables, telegraphs and telephones, burglar and fire alarms, annunciators for hotels or offices, railway signals, lightening rods and

26 Of Koenig's exhibit, Judge Barnard wrote that "there is no other in the present international exhibition which surpasses it in scientific interest". Judge Henry described Bell's telephone in considerable technical detail, concluding that "this telephone was exhibited in operation [...] and was considered by the Judges the greatest marvel hitherto achieved by the telegraph. The invention is yet in its infancy, and is susceptible of great improvements". USCC, *International Exhibition*, op. cit. (n. 5), v. 7, pp. 343 and 452–453. For a fascinating autobiographical account of his performance, see Bell to Mabell Hubbard, 21 June 1876: Washington (D.C.), U.S. Library of Congress, Alexander Graham Bell Family Papers, MSS 51268, box 35. Cf. Robert V. Bruce, *Bell: Alexander Graham Bell and the Conquest of Solitude*, Boston, 1973, pp. 188–198.

magnetic compasses presented in that class. Third, America's industrial production of clocks attracted special comment from the judges. Mass-produced with interchangeable parts, made in "every style and size", and surprisingly durable, these clocks provided yet another instance of the 'American style' at the Philadelphia exhibition. Most distinctive to the judges was the low price of these clocks, so low "as to open for them a market in every country of the globe". American clock manufacturers compete with each other not via quality or innovation, concluded the judges, but primarily via price.

In their citations for individual makers, the judges also repeatedly mentioned value. The teaching collection of H.J. Harting Bank of Utrecht was "more remarkable for its low price than its perfection". Alois Kreidl of Prague was commended "for a very low-priced manufacture of instruments of illustration in science teaching" as was James W. Queen & Co. of Philadelphia for "extremely low prices". An aneroid barometer by Lion & Guichard of Paris drew praise (not criticism) for its "cheap construction". Similarly the Spectacle Makers' Society of Paris displayed spectacles and mathematical instruments of "extremely moderate" prices. The London maker Henry Crouch's student microscope was "remarkably cheap" as was R. & J. Beck's "economic microscope [that] provides for most ordinary investigations in pathology and natural history at a very insignificant expense".[27] Apparently, pecuniary considerations were never far from the minds of the judges, all significant consumers themselves of instruments.

Even more frequently, the judges praised exhibitors for the "excellent workmanship", "good design", "great accuracy", "fine construction", or the "perfection" of their apparatus. A rhetoric of quality pervades most of the citations, as can be seen in a randomly selected set of examples: "commended for excellence of workmanship" (Western Electric Manufacturing of Chicago for its galvanometers); "philosophical instruments for instruction [...] of good quality" (Jacob Pick of Warsaw); a saccharometer and improved hair hygrometer "of excellent construction" (Hermann & Pfister of Berne); a self-registering anemometer "of excellent workmanship" (Theorell & Sörensen of Stockholm); "commended for excellent construction of philosophical apparatus" (G.A. Schultze of Berlin); "commended for accuracy of graduation and [...] general excellence of workmanship" (steel rules of Eagle Square Manufacturing Co. of S. Shaftsbury, Vermont); one of the largest scientific instruments on display, an eight-foot equatorial telescope made by C. Fauth & Co. of Washington, with a

27 USCC, *International Exhibition*, op. cit. (n. 5), v. 7, pp. 491–504. See Henry M. Leicester, "Mendeleev's Visit to America", *Journal of Chemical Education* 34 (1957), pp. 331–333; Michael D. Gordin, *A Well-Ordered Thing: Dmitrii Mendeleev and the Shadow of the Periodic Table*, New York, 2004, pp. 67–68.

7-inch objective glass by Alvan Clark & Sons, was "commended for excellence in design and construction".[28]

The judges looked for novelty, value for money, and quality construction in the instruments on display at the Philadelphia Exhibition. They complained about a lack of the former; but apparently found well-made, cheap apparatus in many of the vitrines. And although occasionally they might comment on highly decorated items, the judges much more frequently stressed the plain, the simple, and the unostentatious. William Thompson's description of Bell's telephone nicely reflects the tone of austerity in the judges' rhetoric. With his ear pressed to the instrument, Thomson "heard it speak distinctly several sentences, first of simple monosyllables, 'To be or not to be' (marvelously distinct); [...] I need scarcely say I was astonished and delighted [...]. This, perhaps the greatest marvel hitherto achieved by the electric telegraph, has been obtained by appliances of quite a homespun and rudimentary character".[29] Homespun and rudimentary, not fancy, is how the judges saw many instruments in Philadelphia.

The Photographs

I have found no autobiographical accounts by ordinary (excluding the panoramic writers) visitors to the 1876 Philadelphia Exhibition that particularly recall their experiences with the scientific instruments on display. But by the late 1860s, photographers had begun to create images of the world expositions. Both the Parisian *Exposition universelle* of 1867 and the Viennese *Weltausstellung* of 1873 were captured in photographs. Not surprisingly, the Philadelphia Exhibition would be the most photographed world's fair to date, and some of the surviving silver albumen prints allow us to see some of the instruments there exhibited.[30]

Given the massive size and complexity of the exhibition halls in Philadelphia, planning officials had issued a set of "Rules and Information for Exhibitors" that stipulated parameters for all displays. The organizers wanted a uniform 'look' for the exhibition. Hence, nothing could extend higher than fifteen feet from the floor or hang from the ceiling (without special permission); nothing could block the aisles; print materials and samples could be tastefully displayed but not forced upon visitors (*Fig. 6.3*). More importantly for our purposes,

28 USCC, *International Exhibition*, op. cit. (n. 5), v. 7, pp. 489–493.

29 Ibid., v. 7, p. 453.

30 The Free Library of Philadelphia holds a large collection of 1876 photographs, available for browsing at libwww.library.phila.gov/CenCol/index.htm (accessed 10 January 2014).

PRICE LIST.

Solar Compass with tripod, ... $220.00
Patent Solar Transit, complete, ... 241.00

Surveyors' Compasses with Staff Mountings.

Plain, ...4 in. $30.00. 5 in. $35.00. 6 in. $40.00
Vernier, ... " 40.00. " 45.00. " 50.00
Railroad, single vernier, ...5 in. $65.00. 5½ in. $70.00
Railroad, double vernier, ... " 80.00. " 80.00
Compass Tripod, $8.00. Jointed, $13.00.

Pocket Compasses with Folding Sights.

2½ inch, $ 9.00. With Staff Mountings $11 50.
3½ " 11.00. " " " 13.50.
" two levels, 15.00.
3½ " Vernier, with staff mountings and levels, $18.00
4½ " " " " " 20.00
3½ " Railroad, " " " 25.00
Miners' Compasses or Dipping Needles,
for tracing iron ore, from $10.00 to $15.00

Transits.

Vernier Transits with compass tripod,
4 in. $75.00. 5 in. $80.00. 6 in. $85.00
Single Vernier Surveyors' Transits, leveling tripod,
4 in. $135.00. 5 in., and 5½ in., $140.00 each.
Double Vernier Surveyors' Transits, leveling tripod,
4 in. $160.00. 5 in., 5½ in. $165.00 each.
Double Vernier Engineers' Transits, leveling tripod,
4 in. $175.00. 4½ in. and 5 in. $180.00
New Miners' Transit, patent solar attachment,
and patent extension tripod, $280.00

Leveling Instruments.

Engineers' Y Levels, 16 in., 18 in., 20 in. and 22 in.,
with tripod, $135.00 each.
15 inch Dumpy Level, with tripod, ... $75.00
11 inch Architect's Level, with tripod, ... $35.00

All kinds of Engineers' Instruments to order.

Send for MANUAL with full price list and description.
Price fifty cents, post-paid.

0777

FIGURE 6.3 *Gurley's trade card for the Philadelphia Exposition, 7 × 13 cm, recto and verso* (AMERICAN ANTIQUARIAN SOCIETY, EPHEMERA TRADE 0777. COURTESY OF AAS).

exhibited items could be placed either directly on the floor, on platforms surrounded by low railings, in individual or compound vitrines, or in inhabitable displays (rooms through which visitors could walk). Judging from the extant photographs, most of the instrument makers in Group 25 choose to place their items in glass-covered vitrines, reflecting practices for the public display of small items that were widespread in late nineteenth-century museums or retail stores.[31]

With space at a premium, the exhibitors of scientific instruments crammed together their products, usually filling every available nook and cranny of the

31 USCC, *International Exhibition*, op. cit. (n. 5), v. 1, pp. 121–122; Giberti, op. cit. (n. 5), pp. 118–139.

FIGURE 6.4 *Rudolph Koenig's Philadelphia exhibit*
(PRINT AND PICTURE COLLECTION, FREE LIBRARY OF PHILADELPHIA, CEDC 021854. COURTESY OF FLP AND BRIDGEMAN ART LIBRARY).

vitrines. Entirely removed from any context that might illuminate their use or function (I have found no evidence of fully dioramic presentations), with few if any explanatory labels or text, the instruments standing alone had to tell the truth about themselves. A nearby attendant might have provided verbal commentary for visitors. But for unattended exhibits, the uninitiated Philadelphia visitors would have been on their own, gazing at scientific apparatus that might have seemed to many quite strange, mysterious, or even unfathomable.

The New York maker of mathematical instruments, Keuffel & Esser, placed three table-like vitrines in its small 'room' (*Pl. 7*), anchored in the back by a neo-classical wall with Doric columns at the sides and a pediment above. Completely covered with large T-squares and other architectural drawing tools, the wall projects an ethos of order, rationality and symmetry into Keuffel & Esser's crowded vitrines, with the company logo prominently featured at the top of the central axis. By contrast, the larger vitrines of Rudolph Koenig are quite unadorned, with simple boards featuring the maker's name attached at the top (*Fig. 6.4*). On the other hand, Koenig's display supported performances. A squat, square bellows stands in front between the two large vitrines, poised to drive air through an aluminum siren sitting on the counter between the vitrines. Between the siren and bellows are stacked several poster-sized boards onto which are affixed paper strips showing manometrically produced graphical representations of sounds, a kind of optical acoustics that Koenig had invented. Visitors surely could have understood representations only with explanations from attendants. Indeed, Koenig's exhibit must have required the continual involvement of attendants to enable visitors not only to see but also to hear the apparatus at work.[32]

The New York mathematical instrument maker, James Prentice, placed his medley of surveying and drawing instruments in a tall vitrine of curved glass and medallions, evocative of the style of Louis XVI then popular among American furniture makers (*Fig. 6.5*). An American eagle, posed to dive, sits above the vitrine. Located just to the left of the Keuffel & Esser exhibit, Prentice's vitrine seems to dwarf the small apparatus it contains. Another mathematical maker, the Stanley Rule & Level Co. of New Britain, Connecticut, also chose a vertical vitrine in a vaguely Louis XVI style, and completely covered its vertical display surfaces with rulers and other mostly flat objects (*Fig. 6.6*). In both of these examples, the vitrine and the instruments come together into a kind of still life; no motion or performance is signaled by these displays.

In one of the larger exhibits of Group 25, the American maker, Howe Scale Co. of Rutland, Vermont, placed literally hundreds scales in a fenced-off area. Large freestanding industrial scales surround vitrines filled with small balances (*Fig. 6.7*). The large maker of both chemical and physical apparatus, Lenoir & Forster of Vienna, also exhibited its wares in a freestanding arrangement, placed on a low table. Chemical glassware, bottles of reagents, and a vacuum pump are crowded together, forming a tableau somewhat reminiscent of a chemical laboratory bench (*Fig. 6.8*). Of all the photographs I have seen of scientific instruments on display at the Philadelphia Exhibition, the exhibit by Lenoir & Forster is the most dioramic, recreating the habitat in which one might expect to find these apparatus in their 'working' lives.

As a final example of instruments and photography in Philadelphia we turn not to images of instruments on display but to photographs on display that featured instruments. In addition to manufactured products, agricultural commodities and fine arts, the Philadelphia exhibition featured numerous displays by America's education institutions, from primary and grammar schools through colleges and universities. Arranged by state, the schools

---

32 Cf. David Pantalony, *Altered Sensations: Rudolph Koenig's Acoustical Workshop in Nineteenth-Century Paris*, Dordrecht, 2009, pp. 58–60, 115–119 and 317–326.

FIGURE 6.5 *James Prentice's Philadelphia exhibit*
(PRINT AND PICTURE COLLECTION, FREE LIBRARY OF PHILADELPHIA, CEDC 022066. COURTESY OF FLP AND BRIDGEMAN ART LIBRARY).

showed off their buildings, curricula, textbooks, and specimens of student work. Dartmouth College, one of North America's oldest colleges founded in 1769, chose to represent itself with, among other things, a set of albumen prints depicting its philosophical apparatus, neatly laid out in a lecture hall. Separate photos presented the disciplines of heat, sound, electricity, mechanics, optics,

FIGURE 6.6 *Stanley Rule & Level Company's Philadelphia exhibit*
(PRINT AND PICTURE COLLECTION, FREE LIBRARY OF PHILADELPHIA, CEDC 020795. COURTESY OF FLP AND BRIDGEMAN ART LIBRARY).

FIGURE 6.7 *Howe Scale Company's Philadelphia exhibit*
(PRINT AND PICTURE COLLECTION, FREE LIBRARY OF PHILADELPHIA, CEDC 042310. COURTESY OF FLP AND BRIDGEMAN ART LIBRARY).

FIGURE 6.8 *Lenoir & Forster's Philadelphia exhibit*
(PRINT AND PICTURE COLLECTION, FREE LIBRARY OF PHILADELPHIA, CEDC 021205. COURTESY OF FLP AND BRIDGEMAN ART LIBRARY).

FIGURE 6.9 *Dartmouth College's Philadelphia exhibit, albumen print of its optical instruments, 21 × 14 cm. Dartmouth College Library Special Collections, Photo File, Scientific Apparatus*
(COURTESY OF DARTMOUTH COLLEGE LIBRARY).

and mathematical instruments (*Fig. 6.9*).[33] Such pictures celebrated not the makers but rather the users of scientific instruments. The Dartmouth photos, however, show neither students nor professors; rather, the instruments stand alone, valorized simply by the richness of each disciplinary still life.

Conclusion

Unlike displays of scientific instruments in museums, private collectors' cabinets of curiosities, makers' shops, scientific institutions (e.g., observatories or laboratories), or in printed books or catalogues, the presentation of such artifacts at the international expositions signaled what we might call a new commodification of science and its tools by the mid-nineteenth century. Instruments had become manufactured goods, mass-produced across the industrializing world, traded across national borders, protected by tariffs and trade policies, and worthy of display among the entire array of industrial

33 USCC, *International Exhibition*, op. cit. (n. 5), v. 7, p. 59.

and cultural products of the nations. Makers of these instruments competed with each other for both local and international markets.[34] By 1876, 'scientific instruments' had become a standard category in the taxonomy of material to be presented at world expositions.

In Philadelphia, the scientific (and musical) instruments merited one of the 36 groups into which the 'world's products' had been placed. As such, they attracted some attention from the panoramic writers, generated worry among official delegations assessing their nations' scientific and industrial competitive position, and received from a distinguished set of international judges more sustained scrutiny than had occurred at previous international expositions. And with its ten million visitors, the Philadelphia Exhibition placed more scientific instruments before more viewers than had ever occurred previously.

On the other hand, the distinguished judges were rather unimpressed by the lack of innovation or novelty in the instruments on display in Philadelphia. Perhaps the transport costs or the high American tariffs on instruments imported for non-educational purposes kept the more innovative makers away from Philadelphia? Perhaps the local (American) makers, like the American clock manufacturers, had increasingly decided to pursue low cost rather than innovative designs for their instruments? In any case, the icon of the 1876 Philadelphia Exhibition would be the large Corliss steam engine, laboring each day to power all the equipment in the Machinery Hall, not a scientific instrument.

---

34 Cf. Mari E.W. Williams, *The Precision Makers: A History of the Instruments Industry in Britain and France, 1870–1939*, London, 1994; Anita McConnell, "From Craft Workshop to Big Business: The London Scientific Trade's Response to Increasing Demand, 1750–1820", *London Journal* 19 (1994), pp. 36–53.

CHAPTER 7

Permanent Demonstrations: The Science Teaching Museum at the University of Chicago

Steven C. Turner

Introduction

In considering the ways that scientific instruments are displayed,[1] location and tradition are often important considerations. This was certainly the case in Chicago between the years 1890 and 1950, when science and scientific instrument manufacturing were important sources of both wealth and prestige. During that period Chicago developed a rich and unique tradition of publicly displaying and demonstrating scientific instruments.

Beginning in the 1880s, American scientific instrument manufacture began to move from the cities of the American east coast, where it had traditionally been centered, to the dynamic new manufacturing and transportation hub in Chicago, Illinois. In the following decades, as millions of European immigrants poured into the United States, an important part of these manufacturers' business became producing the science education instruments and supplies needed to educate the growing population. During this period downtown Chicago was graced with the large and impressive collections of scientific apparatus that these companies prominently displayed in their showrooms and demonstration areas.[2]

The presence in the city of celebrity scientists, like the Nobel prize-winning physicist Albert Michelson at the University of Chicago, also brought scientific instruments to the public's attention. In Michelson's case his fame was so great that the progress of his ongoing research was covered in the daily newspapers,

---

1 This paper grew out of a talk given at the XXIX Symposium of the Scientific Instrument Commission, Florence, Italy, October 4–9, 2010. Because of the conference's theme, an exploration of the 'presentation' of scientific instruments, I situated my story in the broader instrument display traditions of Chicago. That approach is retained in the present article.

2 This and the following discussion of science education draw from Steven Turner, "Chicago Scientific Instrument Makers, 1871–1918, Part 1: the School Science Market", *Rittenhouse* 19 (2005), pp. 65–128; Turner, "Chicago Scientific Instrument Makers, 1871–1918, Part 2: The University and Research Market", *Rittenhouse* 20 (2006), pp. 1–64.

 | DOI 10.1163/9789004264403_008

as was the work of anyone associated with him, including even his instrument makers.

And it should come as no surprise that scientific instruments were prominently featured at all of the city's major expositions. One of the most popular exhibits at the 1893 Chicago Exposition was the giant Yerkes telescope, which was displayed inside the main exposition building. Although missing the forty-inch objective lens, which was still being fabricated by Alvan Clark and Sons, the instrument was still impressive—especially since temporary electric motors allowed the huge machine's range of motion to be demonstrated, which was done for visitors in regularly scheduled performances. Similarly, four decades later, at the 1933 Chicago Exposition, a 200-foot high working thermometer became the unofficial icon of the event. The overall effect was to blur the distinction between public science and science instruction. For example, many of the impressive science and technology displays at the 1893 Exposition were purchased, after the Exposition had closed, by the wealthy Chicago industrialist Phillip Armour, for use in his experimental school, the Armour Institute. And the science exhibits crafted for the 1933 Exposition were built with the intention of using them later in a special gallery of Chicago's new Science and Technology Museum.

This was the context in which, in 1930, the University of Chicago instituted a new program for undergraduate students. Widely referred to as "The New Plan", the goal was to revitalize a curriculum that most agreed was becoming stale.[3] There was also great interest in attracting a higher caliber of students and the new plan sought to do this by exposing students to the latest work and thinking in each of the disciplines. Under this arrangement, in their first two years, students were required to take four year-long general courses, in history and culture, social studies, biology and the physical sciences. It is with this last course that this chapter is concerned.

The Physical Sciences Museum

The "Introductory General Course in the Physical Sciences" covered physics, chemistry, astronomy and geology; in the 1930s the enrollment in the course was generally around 650 undergraduate students. These students tended to have had little previous exposure to science and so placing them in traditional laboratories, especially with such a large class, was out of the question. And yet there was broad agreement among the faculty that the course should require

3 Chauncey S. Boucher, *The Chicago College Plan*, Chicago, 1935.

that students see and understand the basic phenomena of each of the sciences. The solution, it appeared, was to put these experiences all in one place—in a museum of science phenomena.

The Physical Sciences Museum was founded in 1931, just in time for the first group of new students. It was largely the creation of Harvey B. Lemon (1885–1965), a physics professor at the University with a deep interest in education (who would later write many widely used introductory college physics textbooks). The museum was initially almost completely unfunded and Lemon had to work hard to even find a place for it. It ended up in some interior rooms in the University High School. The location was inconvenient, being on the edge of campus and several blocks from the lecture room, but the space was large and well lit, and with nearly 3,000 square feet of space, Lemon was ultimately able to install about 125 permanent exhibits and demonstrations.

Few images of the museum survive, and those that do are of poor quality, but they at least show us how the displays were presented. They were placed sequentially on low tables around the perimeters of the rooms, with instructions and general information about them mounted on the wall behind. Many appear to have been primarily presentations about scientific principles, such as the Bernoulli effect, rather than actual demonstrations.

With essentially no budget, Lemon needed to be resourceful in furnishing his museum. Initially, his displays were constructed from 'existing resources' in the University's physics and chemistry labs. But by 1934 at least half of the displays were on loan from Chicago's Museum of Science and Industry. These displays had originally been exhibited at the 1933 *Century of Progress* Exposition, and had been designed and built with the intention of using them later in a permanent gallery at the Museum of Science and Industry. But construction delays had put that project on hold and Lemon was able to borrow them for the use of his Chicago students.

Over time, a strong relationship developed between the Museum of Science and Industry and the University's science community, with staff from both institutions working collaboratively to design, install and maintain new exhibits. By 1944 Lemon was describing himself not only as a physics professor at the University but also as a curator at the Museum of Science and Industry.

One of the unique things about the Physical Sciences Museum was that it was intended only for the use of students in the Introductory Course and was only open to the public by special arrangement. And while most of the demonstrations were about physics, there were also demonstrations for chemistry, as well as a series of panels for the Geology section. These panels illustrated the Geological Periods and the kind of life they had contained. The astronomy part of the course was taught at the Adler Planetarium, which also provided instruments for the students to study and use.[4]

The Physical Sciences Museum lasted for seven years and the introductory course it supported proved popular with most of the students, particularly the better ones. During this period about forty thousand students and visitors made use of the museum.[5] During the same period, the overall quality of new students at the University increased measurably and there was a widely shared perception that the tone of the campus was being elevated. Perhaps in support of this observation, during the same period the quality of the University's football team dropped precipitously.

But there were problems as well, both with the introductory course and with the museum itself. Particularly in the University's administration, there was criticism of the general courses for being too 'fact' oriented, and particularly in the history and social sciences courses, there was a longing to emphasize the cultivation of 'values' instead of the development of critical thinking. The university president, Robert Hutchins, shared these criticisms and made moves to replace the general introductory courses with a more conventional "Great Books" program.

And there were problems in the Physical Sciences Museum. The Chemistry section of the museum had proven to be difficult and expensive to maintain and there was little support in the Chemistry department to keep it going. In Geology, where an elaborate series of self-contained panel displays had been created, it was found that placing them in the hall outside the Geology lecture room—where students could study them before and after classes—allowed the students to use them more effectively.[6]

The Demonstration Laboratory

So in 1938 the Physical Sciences Museum closed and early in 1939 a new museum, dedicated specifically to physics, was opened in the expansive fourth floor of the Ryerson Laboratory, the University's physics building. As with the original museum, the goal of this new museum was to let students experience

4 The Adler and the University continue to have a collaborative relationship. An advisory committee of science faculty still provides an annual assessment of the planetarium's programs.

5 Harvey B. Lemon, Fitz-Hugh Marshall, *The Demonstration Laboratory of Physics at the University of Chicago*, Chicago, 1939, p. 2.

6 Ibid., p. 96.

the basic phenomena of science. But this time the topic was limited to physics, and Lemon now had a modest budget. The class was still too big and too inexperienced to put in an actual laboratory, but Lemon now used the museum to give them a similar experience.

Widely referred to as the "Demonstration Laboratory," this new museum contained 239 permanent demonstrations. As much as possible, they were built to let the students operate them, but if the apparatus was too delicate it would be automated and connected to a push button. Many, but not all, of the demonstrations were qualitative; student assistants were always present to assist students in operating the apparatus and understanding the lessons. Students could use the lab any time it was open and they were welcome to come back as many times as they wanted.

Although still primarily intended for use by students in the introductory physical sciences course, it now also welcomed students from some of the more advanced Physics courses. The laboratory was also used by students training to be science teachers and once a week it was opened to the general public, although without the benefit of access to lectures or discussion groups, many of these visitors found the experience confusing.

In putting the museum together, Lemon drew on his experience of visiting many public museums, notably the Deutsches Museum in Munich, the South Kensington Museum in London, and the Franklin Institute in Philadelphia. He was impressed with the displays exhibited at the joint headquarters of the National Resource Council and the National Academy of Sciences in Washington, D.C., but his main influence was the Museum of Science and Industry in New York City, only recently opened in 1936.[7] This was the only museum attempting what he called "systematic teaching", which he described as the "logical development and completeness of the science of physics" such as a student might encounter in a formal science course.[8]

Fortunately, in collaboration with Fitz-Hugh Marshall, Lemon wrote a book about the museum that included images of most of the instruments he had on display: *The Demonstration Laboratory of Physics at the University of Chicago* (1939). The following pictures and comments are taken from it. As the layout drawing shows (*Fig. 7.1*), the museum was organized much like a typical introductory physics course, beginning with Mechanics and the properties of matter. The next section was Heat and Kinetic Theory, then Electricity and Magnetism, then Sound and Wave Motion, then Optics and

7 Robert P. Shaw, "The Progressive Exhibit Method", *American Physics Teacher* 7 (1939), pp. 165–172.

8 Lemon, Marshall, op. cit. (n. 5), p. 5.

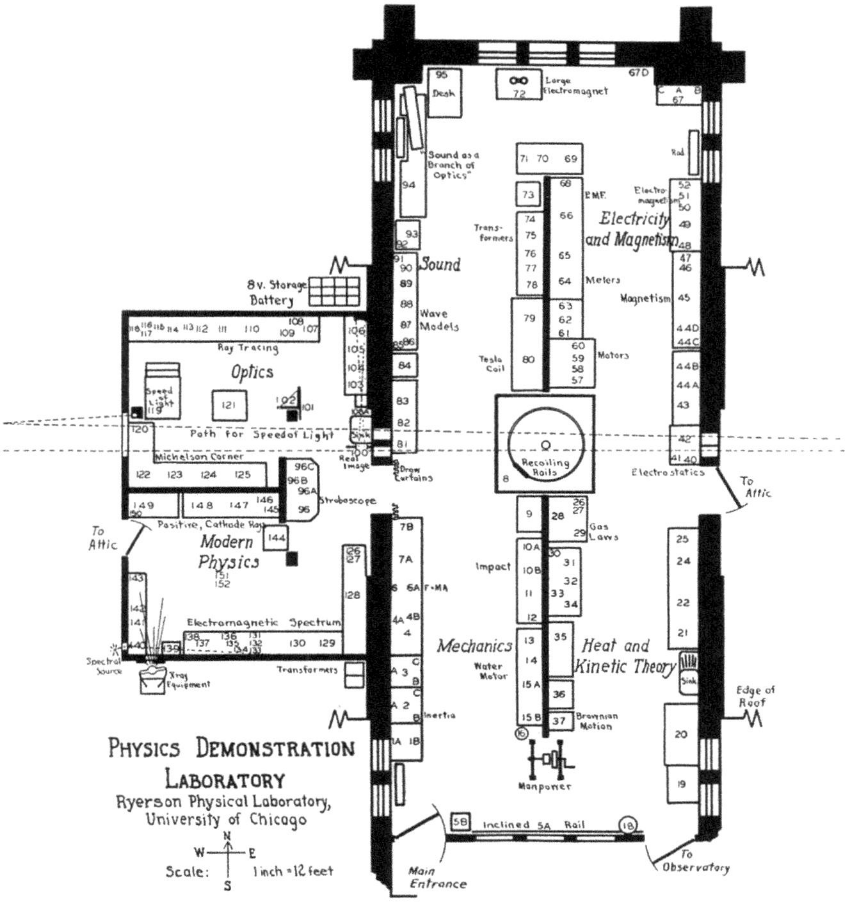

FIGURE 7.1 *Plan of the Demonstration Laboratory. The demonstrations were arranged so that they could be blocked off to prevent students from having access to them before their topics had been covered in the lectures. Students could visit the lab any time that it was open and they were free to visit as many times as they liked*
(HARVEY B. LEMON, FITZ-HUGH MARSHALL, *THE DEMONSTRATION LABORATORY OF PHYSICS AT THE UNIVERSITY OF CHICAGO*, CHICAGO, 1939, P. 9).

finally Modern Physics. Astronomy was now taught in the student observatory, which was conveniently located in an area adjacent to the lab. There was always at least one attendant in the museum while it was open, and usually more. These attendants kept the apparatus working and assisted the students in performing the demonstrations. The only inflexible rule of the museum was that all the demonstrations had to be working whenever the museum opened its doors.

INERTIA

Any motion other than uniform rectilinear motion denotes the necessity of some agent, i.e., the presence of unbalanced forces. This fact is so obvious to anyone that any apparent contradictions (Expts. 1A and 1B) at once stimulate a search for the hidden cause or agent.

1A *2A*

2B

FIGURE 7.2 *The inertia demonstration*
(LEMON & MARSHALL, *THE DEMONSTRATION LABORATORY OF PHYSICS . . .*, CHICAGO, 1939, P. 17).

The first section that students encountered upon entering the lab was Mechanics (*Fig. 7.2*). Here they saw two simple toilet floats suspended from a pipe. One was filled with only air, while the other had been filled with lead. The act of pushing the first one and then the other gave the students a direct experience of inertia, which would lead to a discussion of mass. That concept was further developed as students learned to use first an inertia balance, then a trip scale and then a spring balance (*Fig. 7.3*). There was also a version of the classic demonstration, the guinea and feather tube. In this case, there were two tubes—one filled with air, the other containing a vacuum. Each

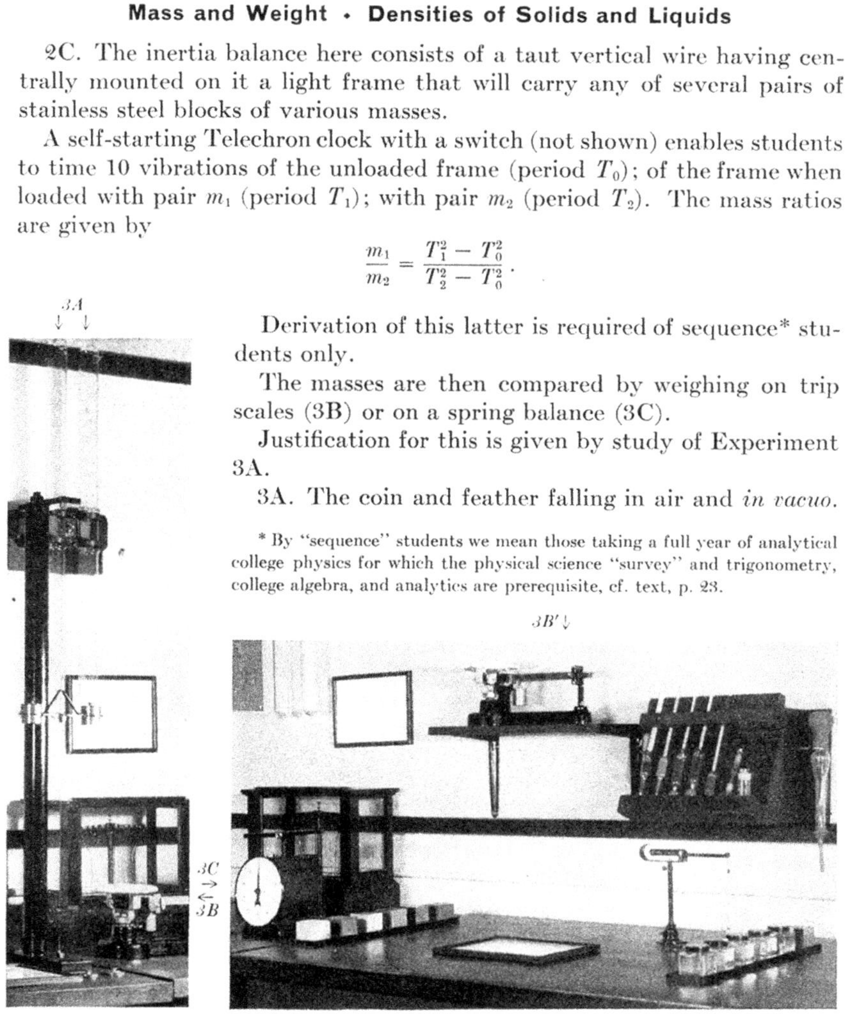

Mass and Weight • Densities of Solids and Liquids

2C. The inertia balance here consists of a taut vertical wire having centrally mounted on it a light frame that will carry any of several pairs of stainless steel blocks of various masses.

A self-starting Telechron clock with a switch (not shown) enables students to time 10 vibrations of the unloaded frame (period T_0); of the frame when loaded with pair m_1 (period T_1); with pair m_2 (period T_2). The mass ratios are given by

$$\frac{m_1}{m_2} = \frac{T_1^2 - T_0^2}{T_2^2 - T_0^2}.$$

Derivation of this latter is required of sequence* students only.

The masses are then compared by weighing on trip scales (3B) or on a spring balance (3C).

Justification for this is given by study of Experiment 3A.

3A. The coin and feather falling in air and *in vacuo*.

* By "sequence" students we mean those taking a full year of analytical college physics for which the physical science "survey" and trigonometry, college algebra, and analytics are prerequisite, cf. text, p. 23.

FIGURE 7.3 *Mass and weight*
(LEMON & MARSHALL, *THE DEMONSTRATION LABORATORY OF PHYSICS…*, CHICAGO, 1939, P. 20).

containing a coin and a feather, the tubes were inverted at the same time to emphasize the contrast in the rate that their contents fell. Each station had a framed description of how to use the demonstration mounted next to it and there was also a printed commentary at each station to guide the students in understanding the demonstration. This understanding was carefully reinforced in the lectures, in follow-up discussion groups, and by the attendants in the lab.

UNBALANCED FORCE: GALILEO'S EXPERIMENT ON AN INCLINED PLANE · BALANCED FORCES ON A FALLING WATER DROP

5A. A steel ball, *S*, is released by hand on the tick of a metronome, *M*, beating seconds, and is allowed to roll down an incline. Its position at each subsequent tick is marked, and the distance between consecutive positions is measured with a stick whose length is cut to be equal to the distance covered during the first second. Distances during the second,

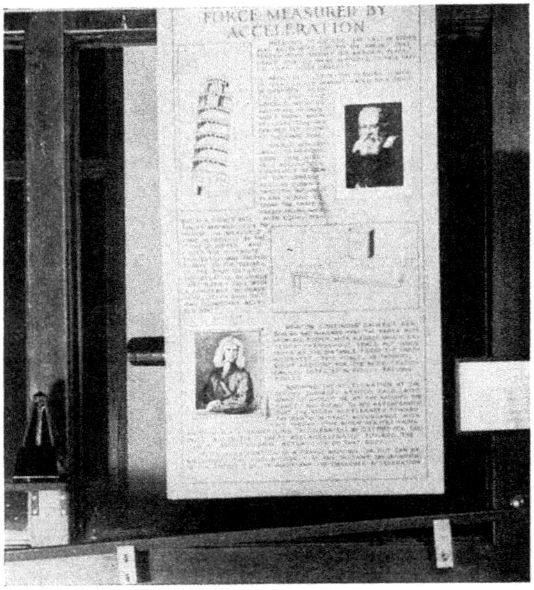

S ←

5A

third, fourth, etc., seconds are found to be three, five, seven, etc., stick-lengths—Galileo's Odd-Integer Law. This law is stated by him in *Two New Sciences*, 2d day, Corollary. A copy of this (English translation by Henry Crew) is on exhibition also. Since the sums of consecutive odd integers are the squares of the integers, the total distance covered in any number of seconds is seen to be proportional to the square of the

5B

24

FIGURE 7.4 *Galileo's inclined plane experiment*
(LEMON & MARSHALL, *THE DEMONSTRATION LABORATORY OF PHYSICS . . .*, CHICAGO, 1939, P. 24).

The lab included a considerable amount of historical information (*Fig. 7.4*). Here Galileo's classic inclined plane experiment is demonstrated, although in this case with a metronome rather than the water-clock that he actually had used. This led to a series of demonstrations about motion, one of which was a Packard's apparatus for showing the acceleration of gravity.

ACTION AND REACTION · CONSERVATION OF MOMENTUM

INELASTIC AND ELASTIC IMPACTS · COEFFICIENT OF RESTITUTION

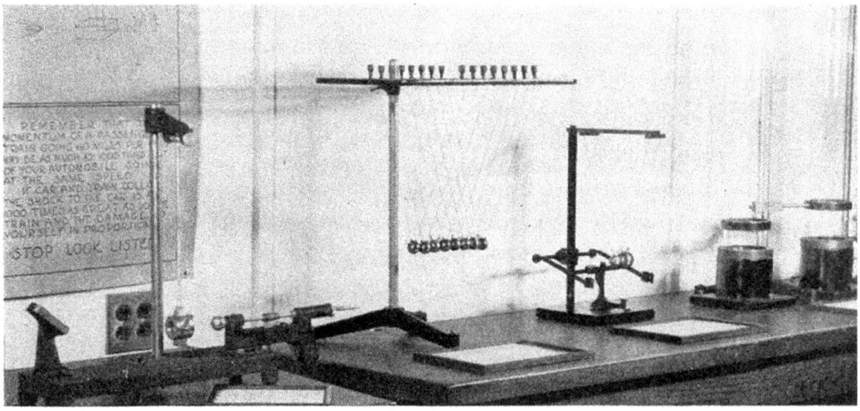

9 10A 10B 11

FIGURE 7.5 *Action and Reaction, and the Coefficient of Restitution*
(LEMON & MARSHALL, *THE DEMONSTRATION LABORATORY OF PHYSICS . . .*, CHICAGO, 1939, P. 28).

Over the course of the school year, as the students' understanding grew, they were given access to more and more sections of the lab, which exposed them to new experiences (*Fig. 7.5*). Here the top image shows a model electric train that is mounted on the rim of a wheel. The wheel could be spun quite rapidly as the train continued on its path. In the center of the lower image is the classic 'Newton's Cradle', used to demonstrate the transfer of momentum. The glass tube on the left was used to drop steel, glass, ivory and lead balls onto a hardened steel block to demonstrate (and measure) rebound height and then to calculate the coefficient of restitution.

Elaborating the force components was an important topic in the course and this section included two more 'inclined plane' demonstrations—a wedge to show 'mechanical advantage' and a typical classroom device to explain the mediating effect of a ramp. Pendulums were used to illustrate the proportionality of force to inertia, and also the transformation of potential to kinetic energy.

I was surprised to find that Lemon considered it important to have a few 'trick' demonstrations in his lab to keep the students on their toes. The one on "perpetual motion" used a carefully concealed jet of air to keep the apparatus in the bell jar moving (*Fig. 7.6*). The laboratory attendants were taught to suggest several improbable possibilities when students asked how it worked—including that the 'nines' were heavier than the 'sixes'. There were several other 'trick' demonstrations in the lab, particularly in the optics section, where virtual images were used to produce some surprising illusions.

The properties of gases and pressure were also explored (*Fig. 7.7*). In the demonstration on the upper right, four different barometer tubes were filled with different gases and the effects of heating and cooling them were demonstrated. In the lower image, demonstrations of Boyle's Law and the principles of buoyancy were presented.

Of special interest in these images are the push buttons that were mounted on the tables and used by the students to activate the displays. Lemon paid particular attention to optimizing the 'push button', or as he called it, the "snap-back switch", and he tested as many as twenty different commercial switches before he developed a way to modify one of them for use in the lab. These customized switches were cheap and reliable, and Lemon insisted that they always work. If a switch did go bad during a demonstration the student assistants were trained to be able to replace them within three minutes—something on which they were tested during their training.[9]

In the Electricity section (*Fig. 7.8*) a series of electrostatic and magnetic demonstrations were presented. At the far left is a solenoid for magnetizing a steel bar. In the center is a 'De la Rive's Tube' (or the 'electric egg' as it was sometimes called) which showed a spark discharge. This was followed

9 Ibid., pp. 116–118.

FIGURE 7.6 *"Perpetual motion"*
(LEMON & MARSHALL, *THE DEMONSTRATION LABORATORY OF PHYSICS . . .*, CHICAGO, 1939, P. 32).

by Oersted's experiment and at the end of the table is a Wimshurst static electric machine that was connected to an 'electric wind' demonstration and could also provide the 'lightning' for a lightning rod display. This was a very popular demonstration and often appears in photos of the museum. The various effects of current electricity were also an important part of any physics course at this period and the lab devoted considerable space to this topic as well.

EXPERIMENTS WITH GASES · VAPOR TENSION AND HYGROMETRY

23. *Vapor tension (upper right).* Four barometer columns contain at the top: mercury vapor only, water, alcohol, and chloroform vapors. The vapor pressures of these substances at room temperature are shown by the depression of the columns below normal barometric height.

Open heating coils are wound on each tube. On closing the switch, the different rates of increase of the different vapor pressures with rise of temperature become apparent. → 25

25. *Hygrometry (upper right).* Human hair and wet-dry bulb hygrometers are mounted in front of a fan for comparative observation of the humidity of the room. →

↓33 2.↓ 29↓ 23↑

FIGURE 7.7 *Experiments with gases*
(LEMON & MARSHALL, *THE DEMONSTRATION LABORATORY OF PHYSICS . . .*, CHICAGO, 1939, P. 35).

FIGURE 7.8 *Electrostatics*
(LEMON & MARSHALL, *THE DEMONSTRATION LABORATORY OF PHYSICS . . .*, CHICAGO, 1939, P. 42).

In addition to motors and galvanometers, the lab also contained a powerful electromagnet, used to give students a direct experience of how magnetic fields affect different kinds of matter. Predictably, there was a demonstration where students were supposed to try to pull an iron pipe off the magnet—which could not be done. An effective demonstration about the effects of differences in metal content could be performed with Canadian and American nickels. The Canadian coins had a high nickel content and could be made to do several amusing tricks when dropped on the magnet. U.S. nickels, by contrast, had a very low nickel content and would barely adhere to the magnet even when placed directly on it.

Although probably too advanced for the introductory students, the lab also contained a corner devoted to the work of Albert Michelson. Michelson's lab had been in the same building and much of his apparatus was still there when the museum was being assembled. In the corner is his first ruling machine for making diffraction gratings and there were several different types of interferometers that he had invented. Remarkably, parts of his 'speed of light' apparatus were also presented and were set up so that students could actually make a rough measurement of the speed of light. To a modern museum curator, the idea of giving students access to instruments of this complexity and historical importance seems almost unimaginable, but their use in the lab may serve to show the commitment of the Physics Department to the success of this project.

The Demonstration Laboratory was surprisingly comprehensive but, impressive as it was, it faced an uncertain future almost from the moment it opened. Although it had other uses, the museum was inevitably linked to the General Courses that Robert Hutchins, the University's new President, opposed.

There were attempts to publicize the museum through newspaper articles and in his capacity as president of the American Association of Physics Teachers Lemon tried to rally support for the museum. But in 1939, just a few months after the Demonstration Laboratory opened, Lemon had a dispute with President Hutchins and abruptly resigned as chairman of the physical sciences general course. This effectively ended his association not only with the course, but also with the museum—which closed shortly thereafter.

I have been unable to find any discussion of the source of Lemon's dispute, but there can be little doubt that the museum's closing was related to the growing crisis in Europe. Germany had just invaded Poland and there was a widespread belief within the American scientific community that war was inevitable. Given the important role that the University of Chicago—and in particular the Ryerson Lab—would soon play in the American war effort, it is not hard to see why having large numbers of undergraduate students and visitors clambering daily up and down the steps of a top secret research installation would be seen as unacceptable.

Although Lemon never attempted to build another museum, he retained an interest in science demonstrations and after the war devoted himself to the new technology of sound motion-pictures, a medium in which he had a longstanding interest.[10]

In 1944 Lemon published a new physics textbook that, although it contained no direct references to museums, did offer large numbers of images taken from a series of physics education films he had made, and which were commercially available.[11] Although the images were small and grainy by modern standards, they (and the films from which they were taken) showed demonstrations of real physical phenomena, many of which were notoriously difficult to perform in a live setting.

Lemon's work with the new technology of 16 mm film, and his use of it to make 'permanent' demonstrations to supplement his textbook, provides an interesting analogy with modern efforts to use digital video and animation for 'permanent' demonstrations on the web. Although Lemon would almost certainly have been fascinated with this versatile technology, given his interest in science pedagogy it seems likely that he would have criticized many current efforts for their lack of connection to a 'systematic' curriculum.

10 Lemon, "The Instructional Value of Certain Types of Motion Pictures", *Physical Review* 19 (1922), pp. 260–261.

11 Harvey B. Lemon, Michael Ference Jr., *Analytical Experimental Physics*, Chicago, 1944.

Conclusion

The University of Chicago's active experiment with science teaching museums lasted only about ten years. But during that time thousands of students successfully used that venue to gain a direct experience of science that would have otherwise remained inaccessible to them.[12]

Although the Chicago experiment was discontinued, it was widely admired and in the 1940s several American colleges and even some high schools started demonstration museums of their own.[13] These museums seem to have generally been effective teaching tools, but typically suffered from insufficient funding and always required considerable personal energy from at least one staff member.

Although the story of the Demonstration Laboratory has been largely forgotten, Lemon's work may survive in modern museums in the strict distinction between 'formal' and 'non-formal' science learning. Formal learning is seen as the systematic presentation of scientific concepts and understanding and is now almost universally seen as properly belonging only in the science classroom. Non-formal learning, on the other hand, is what Lemon would have seen as inspiring or broadening an interest in science and this is now seen as the proper goal for museum science presentations.[14]

---

12 Lemon referred to the "rather unexpected success" of the teaching museum: Lemon, Marshall, op. cit. (n. 5), p. 2. For other evaluations see John W. Boyer, "A Twentieth-Century Cosmos: The New Plan and the Origins of General Education at Chicago," *The University of Chicago Record* 41, (2007), p. 10.

13 Cf. Amherst College's Snell Museum of Physics, which closed in the early 1950s. Lemon encouraged similar efforts by including an Appendix at the end of his book for "those readers who may desire to initiate similar enterprises": Lemon, Marshall, op. cit. (n. 5), pp. 3 and 90–110.

14 See for example: Turner, "Presenting Science Experiences on the Web: Non-Formal Learning in a Museum Context", in Peter Heering, Stephen Klassen, Don Metz (eds.), *Enabling Scientific Understanding through Historical Instruments and Experiments in Formal and Non-Formal Learning Environments*, Flensburg, 2013, pp. 251–265; Sonsoles de Lacalle, Angellah Petruso, "The Value of Partnerships in Science Education: a Win-Win Situation", *The Journal of Undergraduate Neuroscience Education* 11, 1 (2012), pp. A97–A105.

CHAPTER 8

The Display of Twentieth-Century Instruments at Humboldt State University

*Richard A. Paselk**

Introduction

This paper discusses how a number of challenges are addressed in operating and maintaining a small university museum of scientific artifacts. The first challenge is justifying the collection's very existence. Why should this collection exist or be on display? What are the motivations to save and display these artifacts? What unique and/or culturally important information do they represent? Over time the museum's collection and exhibits have been tied into the Humboldt State University's history and development as an important regional center of science and natural resources education; a related extensive web site has made the museum a resource for mid-twentieth-century instrument history.

The Genesis of the Museum

There has been much discussion about the reason for museum collections, particularly at universities, as addressed for example by Richard Kremer.[1] In his article Kremer describes university collections as often beginning as personal collections or hoards of artifacts by university employees. A local 'guardian angel' may then consolidate and organize the aggregations into more formal collections. The collection may then become more museological as artifacts are documented, cleaned, conserved and cataloged. Finally the collections can gain a degree of permanence with the acquisition of funding for storage and display space, staff and the recognition by other university bodies.

* I thank my wife, Gail Paselk, for her careful reading, comments and editorial help on earlier drafts and illustrations for this paper, and Steve Turner for reading and commenting on the subsequent draft.

1 Richard L Kremer, "A Time to Keep, and a Time to Cast Away: Thoughts on Acquisitions for University Instrument Collections", *Rittenhouse* 22 (2008), pp. 188–210.

 | DOI 10.1163/9789004264403_009

Interestingly, the development of the museum at Humboldt State University (HSU) at Arcata, California, has followed Kremer's analysis rather closely. First, it began as a result of collecting, hoarding, and 'dumpster diving' (rescuing from trash containers and trash piles), with the curator keeping/collecting whatever was appealing or 'cool' for the author's personal collection. Second, much of the initiating museum collection came as the result of the dispersal of two campus hoards, one held by the College machinist/electronics technician, and the second held in a Physics Department storage area. Both hoards were 'cleaned out', in 1986 and 1989 respectively, by the University due to space reallocation with unwanted objects going to landfill. Fortunately, the author was made aware of each situation and was able to rescue many items, though many were let go because of the overwhelming number and volume of artifacts—there was no museum yet, and no storage outside of the author's home, garage and on-campus faculty office until the on-campus museum was built in 2000.

About this time the growth of this personal collection initiated the author's thoughts about the direction this personal collection should take—there was too much potential for growth, so it was resolved to focus on chemistry.[2] However, this focus didn't last long as a new collection theme became apparent with the sudden wealth of artifacts from the physics and shop hoards. The new wealth created a potential collection of instruments illustrating the history of science teaching at HSU from its establishment in 1913 through the third quarter of the twentieth century.

While researching and documenting this collection, including items obtained from Humboldt State, starting in the mid-1990s, a real turn came with the development of the World Wide Web and its implementation at HSU around 1995. The timing was propitious, as the author had just completed involvement as Principle Investigator in a seven-year series of educational multimedia development projects and was looking at the web as an educational tool. After years of experience as a developer, constantly requiring the learning of new software and watching key development tools become obsolete, the author was looking for a development platform for his teaching where time could be spent on content, not software. It was already clear that the

2 Ultimately this led to the author's collection of refractometers largely obtained via eBay to enable close observation and manipulation of a variety of these instruments. Because of the isolation of HSU (a six hour drive to the closest large airport or major research university: the University of California, Berkeley) it was impractical to 'visit' instruments elsewhere for the most part. The eBay and HSU instruments provided material for the museum's temporary exhibit, *The Chemist's Refractometer, 1890–1980*, from 2003 to 2005: http://humboldt.edu/scimus/RefracExhibit/Index.html (accessed 10 March 2014).

worldwide use of the web and thus html by hundreds of millions and then billions of users would make html a very stable platform, and that has proven to be the case. Early web pages can still be read and used—html evolves and improves, but it does not go out of date.

To encourage faculty to take advantage of this new technology, HSU established a web development support office, manned mainly by students to help faculty with web projects. The author took early advantage of this situation by first posting course notes and materials on the web for his chemistry courses. The creation of the author's first web museum of university artifacts quickly followed, though the items themselves remained part of a personal collection. Much of the information collected by the author concerning the Humboldt State artifacts in the author's personal collection, originally placed in Word files, was quickly transferred by copy-and-paste into html. Photographs taken earlier on 35 mm film were scanned, digitized and inserted onto instrument web pages. Trade-catalogue descriptions of the instruments were scanned whenever available. After requesting and receiving copyright permissions to post these descriptions from the appropriate makers and vendors, they were also inserted. As a result of these efforts and with help from the campus web development support office, the first web-museum was posted in the spring of 1997. This was the origin of the HSU Chemistry Department Scientific Instrument Museum.[3] However, the 'real' museum, with the artifacts in a university rather than a personal collection, would wait until 2000, as described below.

The HSU museum was fortunate early on to have strong moral support from two successive Deans of the College of Natural Resources and Sciences at HSU who shared an interest in the University's heritage and public image. In 1998, Dean James Smith gave the museum (on-line only at this time) official recognition as a Chemistry Department museum and appointed the author as Curator (new additional title, no new benefits or salary, but it recognized that the museum had academic value to the University and to the author's professional development as a Professor in the Chemistry department). Smith's interim successor, Dean Steve Carlson, discovered the on-line museum and encouraged the creation of a physical museum. He also suggested locating the latter in the university library and negotiated with the Library Dean to arrange it. This was followed by collaboration with the Library faculty and our Art department's "Art Museum and Gallery Practices" faculty to design the museum, followed by case construction and installation by the HSU campus cabinet shop. To fund the cases the author and his spouse donated a sum from an

3 The current museum, now renamed The Robert A. Paselk Scientific Instrument Museum, may be found at http://humboldt.edu/scimus/ (accessed 10 March 2014).

FIGURE 8.1 *Main display cases of Robert A. Paselk Scientific Instrument Museum located in Humboldt State University Library.*

inheritance and received a match from his brother (encouraged by the naming of the new museum after their father, Robert A. Paselk). The museum was officially dedicated October 13, 2000 (by unplanned coincidence, our deceased father's birthday). To accompany the new museum the author completed a major redesign and rework of the entire web site, posted to the web shortly after the dedication (Fall 2000).

The stability of html and the limited resources of the museum (very limited budget, a single individual—the author—to accomplish all coding, research, writing, photography, conservation, cataloging, exhibit design and installation) also recommended it as a single platform for all inventory and other functions of the web museum. The only software development tools currently used at the museum are Adobe Dreamweaver for coding and editing html, Adobe Photoshop for image preparation and editing, and MS Word for writing text, and creating exhibit labels and signs.

Does a local museum like the one at HSU have a rationale beyond the passion of its founder? The collection does not include historically important or valuable objects for the history of science on a national or international scale; it focuses on instruments used in science instruction at Humboldt State. Thus a significant justification of the collection is to tell a local story of science education and the establishment of Humboldt State as an important educational institution in the natural resources and sciences during the mid-twentieth century. The core collections and displays, both online and physical, reflect this in their chronological arrangements.

In the main Library display cases (*Fig. 8.1*), the south case (*Pl. 8*) contains instruments from HSU's early period (1913–1935), when it was named Humboldt State Teachers College. The north case (*Fig. 8.2*) presents instruments from

FIGURE 8.2 *Main display cases, Right-hand (North) case.*

HSU's formative period as Humboldt State College, when it developed into an institution with an international reputation in natural resources and science education. As the collection has expanded it has become possible to place satellite displays in buildings where the instruments were actually used and where the display supports a connection between the teaching activities of the present with the use of classic instrumentation and apparatus before the digital age. Two display cases installed in 2007–2008 highlight the balances used in chemistry courses from 1950–1980 (*Fig. 8.3*) and a grouping of high-quality, state-of-the-art instruments used in HSU's physical and analytical chemistry laboratory in the 1960s (*Fig. 8.4*). A case enclosing a Cary 14 UV-Vis scanning spectrophotometer and some accessories is in progress. In 2011 a third case was added to the main Library display area allowing for temporary, thematic exhibits of current interest, e.g. "Early North Coast Scientific Instruments c. 1850–c. 1950" (January 2012–June 2013),[4] or an exhibit focusing on instruments taken by the University's namesake on his Latin American Expedition, "Alexander von

4 The online version of the exhibit still lives on at: http://humboldt.edu/scimus/NorthCoastInst/N_C_InterDis.htm (accessed 10 March 2014).

Humboldt's Latin American Expedition (1799–1804) Instruments, Geography & Botany" (August 2013–August 2014) as part of HSU's Centennial celebrations (*Fig. 8.5*). Clearly, the museum showcases local university history and the practice of chemistry there since c. 1930.

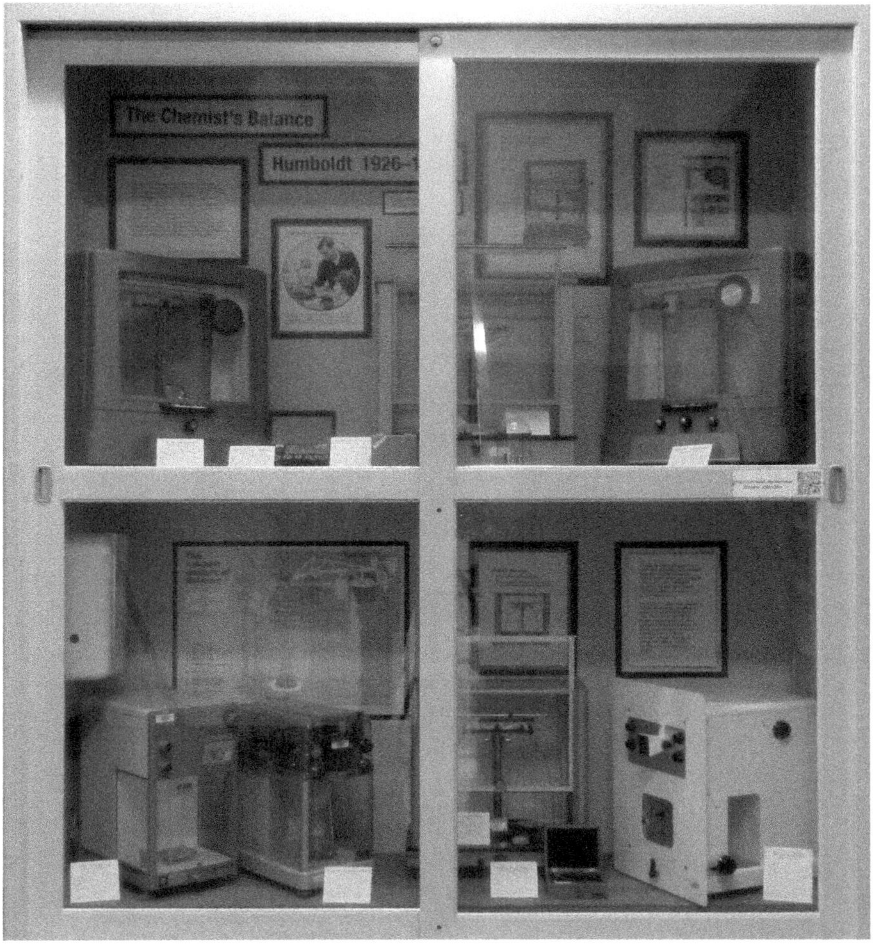

FIGURE 8.3 *Display case for* The Chemical Balance at Humboldt State, 1926–1986. *This case is located near the Chemistry labs where balances are now used, allowing students to see the precursors to our modern digital balances.*

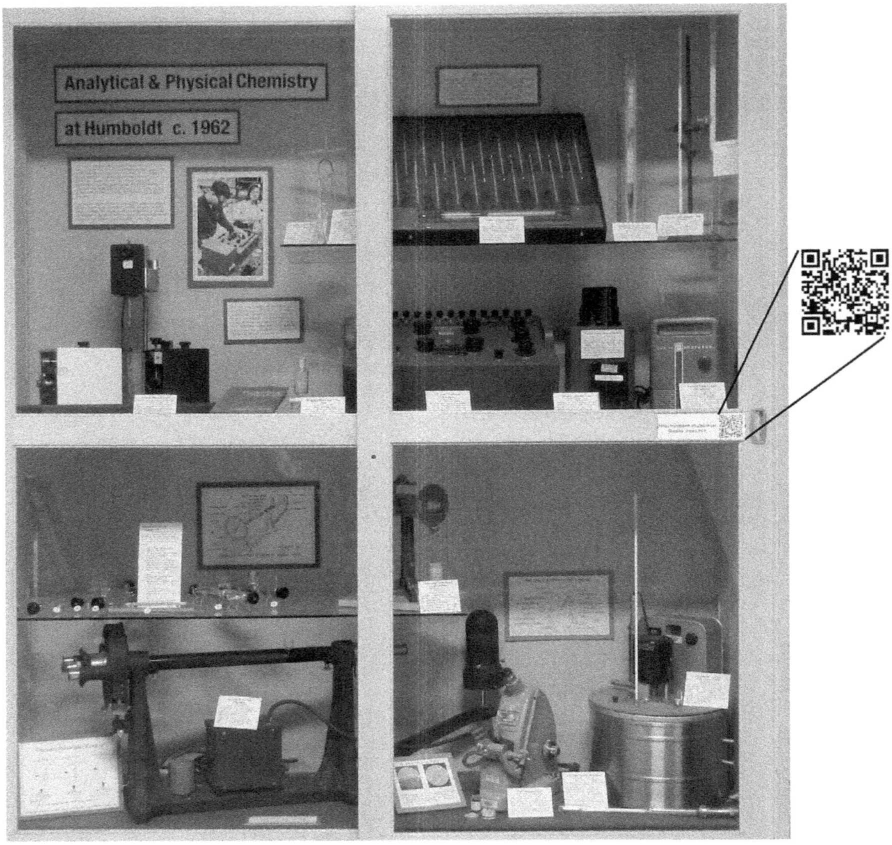

FIGURE 8.4 *Display case for* From the Analytical/Physical Chemistry Lab, c. 1963. *This case is located just down the hall from the two laboratories where this equipment was used. Note the magnified* QR *code that can take the visitor with appropriate technology to the web page for this display.*

Physical vs. Virtual Visitor Experience

A visit to the museum's web site exemplifies the museum's dual web vs. physical nature.[5] The visitor may choose two complimentary paths to explore the museum: a university visitor enjoying the displays may click on the display case pages to access information on the individual displays and artifacts;[6]

5 See the museum's home page: http://humboldt.edu/scimus/index.shtml (accessed 10 March 2014).

6 See: http://humboldt.edu/scimus/Display_index.htm (accessed 10 March 2014).

FIGURE 8.5 *Display case for* Alexander von Humboldt's Latin American Expedition (1799–1804) Instruments, Geography & Botany.

a virtual visitor may click the on-line exhibit pages for greater access to the museum, including objects, documents, a literature collection, some maker histories and web pages for past temporary displays.[7] For public access in the HSU Library,[8] university-maintained computers are conveniently located within the museum area, as can be seen in the upper image on the introductory web page and in *Figure 8.1*.

The university visitor using the display case page may go to specific web pages for any of the museum's display cases by clicking on either the display case titles or case icons. Alternatively, the web site for each display case is designed to enable instant access via hand-held devices (WiFi or 3G) to the case collection documentation for visitors. Each case has a QR code (two-dimensional or matrix barcode) affixed to it, allowing smartphone and iPad or tablet computer users to go directly to the display case web page simply by aiming their QR code-enabled device (*Fig. 8.4*). The display case pictures

7 See: http://humboldt.edu/scimus/Virtual_index.htm (accessed 10 March 2014).

8 All students, staff and faculty have open access. University visitors may obtain temporary passwords to access university computers.

are image-mapped—clicking on the instruments in a display case image takes the user to the specific artifact web page with digital photos of instruments and components, curator discussions and descriptions, and links to trade catalogue descriptions, manuals, brochures and other literature when available.[9] Students and others can not only marvel at the equipment used by their predecessors, they can also explore the use, maintenance and other aspects of these tools of education and discovery, both at the main displays in the library and in displays near the labs where exhibited instruments were once used.

The visitor to the virtual museum can find all of the information provided to the on-site visitor, plus information on additional artifacts in the collection that are not displayed. In addition the curator has provided maker histories, brochures, catalogues, references, and other literature when available such as personal narratives of users and ex-students.

The primary organization of the artifact collection in the virtual museum is in three parts based on the designation of the institution and time period when they were obtained and first used, as seen at the top of the on-line exhibit main page.[10] Clicking on any of these links takes the user to an overview page with links to some institutional history and the instrument collection for each part. These three instrument pages are intended to include all of the museum's catalogued collection of artifacts. Those displayed in the Main Library cases and satellite cases may be alternatively accessed via case display pages. The satellite links take the visitor to the respective Exhibit Catalogues, which include both instruments in the case and related apparatus in the collection plus, particularly for the Balance and Refractometer exhibits, related literature and other documentation. Web pages for Special Exhibits are also maintained and are accessible via this page after the actual exhibits are taken down. Finally, various pages indexing the museum's collections of brochures, essays, and other supporting documentation are linked at the bottom of the on-line exhibit page. All of these web pages have been created and maintained by the author.

---

9 For an example of a typical 'complete' artifact web page visit http://humboldt.edu/scimus/HSC.54-70/Descriptions/Tensiometer.htm (accessed 10 March 2014). Not all instruments are this well documented. Catalogue descriptions, manuals, company histories, and original literature descriptions are often unavailable and not included.

10 See: http://humboldt.edu/scimus/Virtual_index.htm (accessed 10 March 2014).

Americans with Disabilities Act (ADA) Compliance

Recent actions in the State of California have set deadlines for making all online materials accessible to people with disabilities in accordance with national and international standards.[11] The California State University (of which HSU is a part) and HSU's campus also embrace a philosophy ("Universal Design") of making our web pages as usable and 'friendly' as possible to all. In this light I will discuss briefly some of the steps that have been taken to make the web museum more accessible to visually impaired or disabled users. Because the museum was first posted before many accessibility standards were in place, early and updated examples of one of the instrument pages will be compared. Differences that are not visually obvious but which affect users of screen readers will also be described.

In designing the original pages bold and italics as well as font sizes were used to emphasize titles, important points, etc. Unfortunately these signals do not translate well for screen readers (software and/or devices converting written text to audible words) used by visually impaired users. First 'bold' and 'italic' are reimplemented as 'strong' and 'emphasis' respectively, while font sizes are replaced with 'heading' sizes. The heading designations allow sophisticated screen readers to skip from heading to heading much as a sighted user does when scanning a page. These changes do not alter the appearance of the pages, but they make the screen reader experience more convenient and closer to the experience of sighted users.

Another design change is to avoid the use of tables as layout tools. Screen readers for visually challenged users read through the formatting as well as the content for the table as it is read, thus tables are discouraged for formatting. Tables are acceptable for actual tabulated information or data as long as proper table formatting tags are used in the code, an operation built into HTML coding software such as Adobe Dreamweaver™, the tool now used to write the pages for the HSU's web museum. Current pages show some differences, most notably the illustrations and headings, formerly arranged across the page formatted using a one-row table, are now 'stacked' and centered making the pages a bit taller and less compact than in the original. Finally, all images now have 'alt tags'—short descriptions of the images that are read by screen readers to help the user to interact with the images. For 'decorative' images a simple tag is reasonable. Accordingly

11 International guidelines and help for accessible development may be found at www.w3.org/WAI/ (accessed 10 March 2014).

short tags "photo icon" and "catalogue scan icon" are routinely used for the images used as visual 'buttons' on the web site.

Conclusion

In summary the origins and growth of the Robert A. Paselk Scientific Instrument Museum at Humboldt State University closely followed the historical patterns noted by Kremer and listed at the beginning of this article: initiated as hoards and private collections followed by consolidation by a 'guardian angel' and then formalization after recognition by university authorities. This evolution and the desire to create a sustainable museum resulted in the development of a philosophy and reason d'être for the museum, viz., to tell local institutional history, to contribute to university heritage. However, the development of the World Wide Web and its availability at the time of the consolidation of the initial collection also played a major role in the shaping and formalizing of the museum; without the web the museum would likely have remained a private collection to be dispersed by the collector or his heirs. The World Wide Web originated and remains a central focus of the museum. Because of limited time and resources the web display has also become the central record of the museum, with all notes, descriptions, etc. kept online and accessible to all—no other inventory software is currently used. The evolution of the web and society has required on-going redesign to accommodate visually impaired users.

So what does the future hold? The author's career at HSU will end in just over a year, and there is currently no one else involved with the museum. From my dozen-plus years experience with the museum displays, I project that the displays will be stable with no curatorial care for at least a decade. One of my goals over the next year is to try to create a mechanism to sustain the museum and/or find a successor as curator. I may continue maintaining the website for the intermediate future as that can be done remotely, but ideally I will be able to pass this on as well. Time will tell.

CHAPTER 9

Slide Rules on Display in the United States, 1840–2010

Peggy Aldrich Kidwell and Amy Ackerberg-Hastings

Introduction

Reading over the titles of the papers in this volume gives us pause. Others are talking about the grandeur of telescopes, elegant astronomical books, and the dramatic sparks of physical apparatus. Even those of us who are quite fond of slide rules would not claim that they had the visual appeal of many of these objects. For most of its long history, the slide rule has been a tool for use; very rarely was it designed primarily for display.

Nonetheless, three aspects of the slide rule on display that date from the nineteenth century might merit our attention. First, slide rules were among the most widely seen of scientific instruments. From about 1890 until their displacement by electronic calculators in the 1970s, slide rules adorned the desks, lab benches and pockets of workaday American scientists. This reflects both the pervasive role of mathematics in the natural and social sciences and the fact that the instruments were easily portable. Slide rules were produced in large numbers and sold around the world. Second, displays of slide rules at fairs, especially world's fairs, did much to popularize the instrument. While such exhibits did not attract the large crowds that flocked to the electrical displays, they did much to attract the attention of the scientific and engineering press and of instrument dealers. Finally, slide rules were shown at a few public institutions as examples of American ingenuity. Three mid-nineteenth century slide rules in the Smithsonian collections well illustrate these three aspects of the display of the instrument.

In the twentieth century, oversized slide rules for display in the classroom, which had been made earlier in Europe, became common in the United States. The slide rule also gained a modest place in museum exhibitions on innovation, on the history of computing, and on the lives of scientists and engineers. We shall suggest briefly how this occurred at the Smithsonian Institution in Washington, D.C. With the advent of the World Wide Web, it is now possible to place images of entire collections of slide rules on display. We consider this

 | DOI 10.1163/9789004264403_010

in the final section. Examining objects for this purpose offers both an opportunity for scholarship and an occasion to think about how slide rules were displayed in the past.

Slide Rules for Personal Adornment

Not long after John Napier's discovery of logarithms in the early seventeenth century, instrument makers suggested how logarithmic scales that slid past one another could be used in calculation. Later in the century, the British author Henry Coggeshall published a pamphlet describing a linear rule especially for calculations relating to timber.[1] As one of the major early exports of the United States was timber, it is not surprising that a form of Coggeshall's rule called the carpenter's rule came to be one of the first slide rules sold in that nation. The photography collections at the National Museum of American History (NMAH) include a daguerreotype of an anonymous workman proudly holding a carpenter's rule (*Fig. 9.1*). In the mathematics department of the NMAH there is the rule itself (*Fig. 9.2*). It is marked "S.A. JONES & CO. / HARTFORD-CON." City directories indicate that Solomon A. Jones, rule maker, was in business in Hartford, Connecticut, by 1838 and remained there through at least 1849.[2]

The carpenter's rule is made from two wooden one-foot (30 cm) rules held together at one end by a metal joint. Unfolded, one side becomes a simple two-foot measuring rule. The upper part of the other side of the two-fold rule contains a groove that holds a brass slide, with logarithmic scales on the upper and lower edges of both the slide and the adjacent parts of the groove. The inner edges and lower part of the carpenter's rule are marked with other scales of use to carpenters and spar-makers.[3]

The anonymous carpenter in the daguerreotype displayed his rule on a special occasion—when he was photographed. From the late nineteenth century, computing people in the United States, especially men, actually wore slide rules in the course of routine work. They might be designed to fit in the vest

1 Henry Coggeshall, *Timber-Measure by a Line of More Ease, Dispatch and Exactness than any other way now in use, by a Double Scale*, London, 1677.

2 Kenneth D. Roberts, *Introduction to Rule Collecting*, Hartford (Ct.), 1982; and Robert E. Nelson (ed.), *Directory of American Toolmakers: A Listing of Identified Makers of Tools Who Worked in Canada and the United States before 1900*, n.p., 1999, p. 430. The second reference gives the dates 1838 to 1841.

3 Roberts, "Carpenter's and Engineer's Slide Rules", *The Chronicle of the Early American Industries Association* 36, 1 (1983), pp. 1–5.

FIGURE 9.1 *Daguerreotype of a workman holding a carpenter's rule, ca. 1840* (SMITHSONIAN IMAGE ET2013-39489).

pocket like a pocket watch (*Fig. 9.3*). They might be carried in cases suspended from the belt (*Fig. 9.4*). They also might fit in the shirt pocket—sometimes with a flap to hold them in place. A few military slide rules were designed to be strapped to the leg. A small amount of jewelry was in the shape of a slide rule. For example, machinist, historian, and former Smithsonian employee George Norton donated to the National Collections a tie clasp in the form of a linear rule, with slide and indicator. He purchased it from Edmund Scientific Company, a distributor in Barrington, New Jersey, in the 1960s. We never actually saw him wear it on his tie.

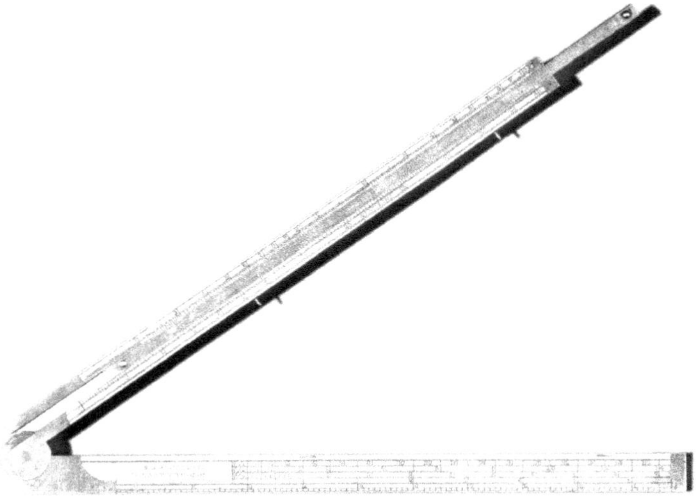

FIGURE 9.2 *Carpenter's rule by Solomon A. Jones & Company of Hartford, Connecticut, ca. 1840*
(SMITHSONIAN IMAGE 2004-40575).

FIGURE 9.3 *Carpenter's slide rule for the vest pocket, ca. 1920*
(SMITHSONIAN IMAGE 95-6863).

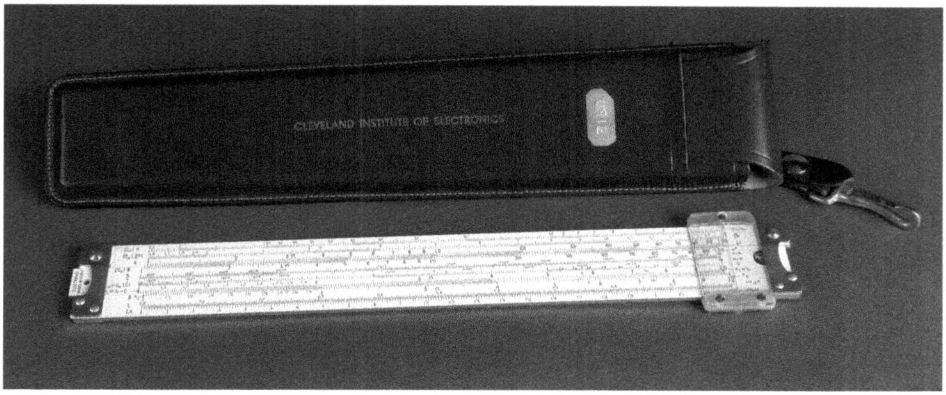

FIGURE 9.4 *Pickett duplex slide rule with case modified to be worn from the belt, ca. 1964* (SMITHSONIAN IMAGE 2009-25619).

The rings, loops, hooks, straps and other fasteners on slide rules offer ample evidence that they were to be worn. Most people did not follow the example of the nineteenth century carpenter in the daguerreotype. Their formal portraits and yearbook images did not include slide rules. Informally, people and their rules might be captured with snapshots. A diverse selection of these images is now available online.[4]

Slide Rules Displayed by Vendors—The Fairs

Slide rules were displayed by vendors as well as by their customers. In the nineteenth century, local, regional and international fairs were important to the distribution of numerous commercial products, from agricultural goods to steam engines to works of art to slide rules. Accounts of these fairs offer important though fragmentary evidence concerning the diffusion of the slide rule.

Promotional efforts of innovators Aaron Palmer and John E. Fuller, who sold one of the first slide rules designed in the United States, illustrate this aspect of instruments on display (*Fig. 9.5*). Palmer apparently was born in Canada around 1810 and settled in upstate New York by about 1830. In 1837, he took out his first patent, for an improvement in endless chain horsepower for driving

4 Some of these images have been posted by Michael Konshak and his associates at the online Slide Rule Museum: http://sliderulemuseum.com/ (accessed 11 March 2014). Others may be found through search engines such as Google Images.

FIGURE 9.5 *A late version of Palmer's computing scale, 1860s* (SMITHSONIAN IMAGE 87-4844).

machinery.[5] He soon became interested in computing devices and, according to later advertisements, spent three years laboring incessantly on his slide rule. In 1842, Palmer, by then a resident of Brockport, New York, published in nearby Rochester *A Key to the Endless, Self-computing Scale, Showing Its Application to the Different Rules of Arithmetic &c*. To garner support for his invention, he solicited the recommendations of local teachers, clergymen, and engineers, publishing these in his *Key*.[6] According to a report in a local newspaper, he

5 Aaron Palmer, "Endless-Chain Horse-Power for Driving Machinery", United States Patent 398, September 22, 1837.

6 Palmer, *A Key to the Endless, Self-computing Scale, Showing Its Application to the Different Rules of Arithmetic &c.*, Rochester (N.Y.), 1842. For historical discussion about Palmer and his associate John E. Fuller see Florian Cajori, "Aaron Palmer's Computing Scale", *Colorado College*

also exhibited his circular slide rule.[7] During the next year, he collected further testimonials from authorities in Boston and Cambridge, Massachusetts; copyrighted the computing scale; published an 11 inch square version of the device in Boston; and exhibited it among the machinery and models shown at the Sixteenth Annual Fair of the American Institute in New York City, held in October. This garnered him a diploma.[8] Palmer continued to improve his device, developed versions in several sizes, took out an 1844 copyright, and exhibited once again at the Annual Fair of the American Institute in 1845. An account of the fair praised the computing scale as "a production which exhibits uncommon genius, and industry, and patience".[9]

Palmer's interests were by no means confined to mathematical instruments. An article on the New York Agricultural State Fair, from 1846, indicates that he exhibited there not only his computing scale but also a looking glass and picture frames.[10] By mid-century, he had sold the rights to his slide rule to John E. Fuller and turned his attention to agricultural inventions.

Fuller had been publicizing Palmer's computing scale since at least 1844.[11] Once he gained rights to the instrument, he set out to improve it. Fuller designed another, non-logarithmic circular scale for computing the number of days between two dates on the calendar. He copyrighted the instrument in

---

Publication (1909), pp. 111–118; Cajori, "John E. Fuller's Circular Slide Rules", *Colorado College Publication* (1909), pp. 120–122; Bobby Feazel, "Palmer's Computing Scale", *The Journal of the Oughtred Society* 3, 1 (1994), pp. 9–17; and Feazel, "Palmer's Computing Scale—Revisited", *The Journal of the Oughtred Society* 4, 1 (1995), pp. 5–7.

7 *Daily Democrat* [Rochester (N.Y.)], August 12, 1842.

8 Palmer, *Palmer's Computing Scale*, Boston (Mass.), 1843; Palmer, *A Key to the Endless, Self-computing Scale, Showing Its Application to the Different Rules of Arithmetic &c.*, Boston, 1844 (copyrighted 1843). On the award from the fair, see "Third Annual Report of the American Institute on the Subject of Agriculture. List of Premiums Awarded at the Sixteenth Fair", in *Documents of the Senate of the State of New-York, Sixty-Seventh Session, No. 124*, Albany (N.Y.), 1844, p. 69.

9 Palmer, *Palmer's Pocket Scale: with Rules for Its Use in Solving Arithmetical and Geometrical Problems*, Rochester (N.Y.), 1845; "The Eighteenth Annual Fair of the American Institute", *The American Review: A Whig Journal of Politics, Literature, Art and Science* 2 (1845), pp. 538–542: 541.

10 "Agricultural State Fair", *Daily Advertiser* [Auburn (N.Y.)], September 24, 1846.

11 *Evening Journal* [Albany (N.Y.)] February 27, 1844, had an advertisement for agents for Palmer's computing scale, as well as a short article on the instrument, excerpted from the Lowell *Courier*. The *Courier* article not only waxed eloquently on the device, but reported that "Through the politeness of Mr. John E. Fuller, we have been shown this valuable work; and were most delightfully surprised in witnessing its operation."

1845, dubbed it "Fuller's Time Telegraph," and arranged for it to be published on the back of Palmer's computing scale (*Pl. 9*). He also prepared a new version of the *Key* to the instrument, which he published in 1846. Here Fuller claimed that he had introduced about 4,000 people to the new scale within the past eighteen months.[12]

In the late nineteenth and twentieth centuries, American inventors of computing devices would be known for selling their products abroad. The cash registers of NCR, the adding machines of Burroughs, and the tabulating machines of Herman Hollerith and later IBM sold not only in the United States but also in Europe and throughout the world. This entrepreneurial vision inspired John Fuller as well. From about 1848 through 1851, he traveled about Europe, demonstrating his instrument in England, Scotland, Ireland, Belgium, France, Germany and Holland. In the course of his travels, he garnered testimonials from such distinguished mathematicians as Augustus de Morgan.[13] He arranged for an English edition of the instrument.[14] He also found a spot among the American exhibitors at the first world's fair, the Crystal Palace exhibition held in London in 1851. Here he is listed as John E. Fuller of Boston, Massachusetts, exhibitor of a "mathematical telegraph and model calculator. Computing telegraph. Pocket computer".[15] This small display received no special commendation from the jurors of class X, which was concerned with "Philosophical Instruments and Processes Depending on Their Use." They were more interested in the Staffel and the Schilt adding machines, the Thomas arithmometer, and the slide rules of Leon Lalanne.[16]

Despite this tepid reception, Fuller was happy to mention his display at the Crystal Palace in later publications on the computing telegraph. The lengthy title to the 1852 edition of the directions for the instrument reads: *Telegraphic Computer, a Most Wonderful and Extraordinary Instrument, by which Business Questions, of Every Possible Variety are Instantly Performed: A Safe and Speedy Check to Avoid Vexatious Errors, Affording at the Same time A Greater Amount of Practical Business Knowledge, than Can Be Obtained for Ten Times the Cost of the*

12 John E. Fuller, *Improvement to Palmer's Endless Self-computing Scale and Key*, New York, 1846.

13 Fuller, *Telegraphic Computer*..., New York, 1852, unpaginated.

14 Fuller, Palmer, *Fuller's computing telegraph multiplication and division by one single process: the great master key rule, by which all problems are instantly solved*...*: to which is added Fuller's complete and perpetual calendar*..., Liverpool, n.d.

15 *Official Descriptive and Illustrated Catalogue of the Great Exhibition of the Works of Industry of All Nations*, London, 1851, 3 vols.; v. 3, p. 1464.

16 *Reports by the Juries on the Subjects in the Thirty Classes into which the Exhibition Was Divided*, London, 1852, p. 310.

Work. This Work Occupied a Prominent Place at the Crystal Palace.[17] In the early 1860s, another version of Fuller's Time Telegraph was published in London and Liverpool. Here, the number of testimonials from satisfied customers is sharply reduced and the references to the Crystal Palace exhibit muted. However, two of the four press reports cited note that the calculating device was exhibited at the Crystal Palace.[18] Thus exhibition at fairs played an enduring role in the sale of this early American slide rule.

Fuller was not the only slide rule maker who sought to benefit from a world's fair. At the 1867 fair in Paris, the Frenchmen Tavernier and Gravet exhibited the one-sided slide rule designed by their countryman Amadée Mannheim. They were commended for building a large demonstration slide rule that could be seen by an entire school class.[19] We shall return to the subject of demonstration slide rules presently. Of more direct importance for the spread of the slide rule in the U.S. was the International Inventions Exhibition held in London in 1885. A linear slide rule made by Gravet alone, as well as cylindrical slide rules of the Irish engineering professor George Fuller and American 'computing engineer' Edwin Thacher, were particularly influential. Both Fuller's and Thacher's rules were initially manufactured by the English firm of Stanley, and were exhibited in 1885 among that company's products. Stanley won a gold medal for its efforts.[20]

The English physicist and instrument inventor Charles V. Boys attended the Exhibition and took the occasion to write an article on the slide rule that was published in New York in Van Nostrand's *Engineering Magazine*. Boys firmly believed that scientists and engineers should use slide rules. He wrote:

> It is a perpetual source of amazement to those who are familiar with this instrument that its use is not almost universal. People of every class have to make simple calculations, while those engaged in scientific work, in designing apparatus, or in invention, perpetually cover sheets of paper with figures, all of which trouble, and the loss of time which it involves,

17 Fuller, *Telegraphic Computer…*, *cit.* (n. 13).

18 Fuller, *Fuller's Computing Telegraph…*, London and Liverpool, n.d. The instrument includes references to slave states and to data from the 1860 U.S. Census, suggesting the date 1861 to 1865.

19 E. Grateau, "Instruments de Mathématiques et Modeles pour l'Enseignement des Sciences", in Michel Chevalier (ed.), *Rapports Jury International*, 1868, 13 vols.: v. 2, pp. 523–543.

20 William F. Stanley received a gold medal at the fair for "improvements in philosophical instruments." See International Inventions Exhibition, *The Awards of the International Juries, Confirmed and Issued by the Jury Commissioners*, London, 1885, p. 32.

> might be saved by the intelligent use of a good slide rule, and yet, for reasons difficult to find out, the habitual use of this instrument is limited to a very small proportion of the calculating community.[21]

Boys went on to describe several of the slide rules exhibited in 1885, placing special emphasis on the advantages of Gravet's linear rules.

In the next decade, the handheld linear slide rule on Mannheim's design spread throughout England and the United States.[22] Instruction in the slide rule joined experiments with far larger machines (e.g., steam engines) in the laboratories of engineering schools. Improved materials, particularly celluloid, made it possible to make relatively inexpensive, sturdy slide rules that were easy to read. British, German and American firms soon were manufacturing slide rules. American makers like Keuffel & Esser displayed their wares at fairs and, if they won medals, pictured them on the title pages of their catalogues.

By the early twentieth century, however, fairs, including international expositions, became far less important for introducing new goods, a task now increasingly achieved by trade literature, published advertisements, and trained sales forces. Moreover, with the slide rule well established in the school and laboratory, general public demonstrations became far less necessary for raising awareness of the instrument.

Slide Rules Displayed by Government Institutions

Nineteenth-century Americans took great pride in demonstrations of their nation's ingenuity. In addition to lauding exhibits at fairs, they took considerable interest in patent models. Models demonstrating that an invention could work were required as part of American patent applications from the early years of the republic. An 1836 patent law authorized a new Patent Office building, with accommodations for staff and extensive public galleries for viewing models. In December of that year, the old quarters of the Patent Office burnt to the ground, destroying both patents and models. Despite this inauspicious beginning, the operation of the reorganized Patent Office proceeded. Would-be inventors, patent agents and the general public soon could and did visit the

---

21 Charles V. Boys, "The Slide Rule", *Van Nostrand's Engineering Magazine* 33 (1885), p. 512.

22 "Gravet's Calculating Rule", *Engineering* 44 (1887), p. 512; Bob Otnes, "Keuffel & Esser—1880 to 1899", *Journal of the Oughtred Society* 10, 1 (2001), pp. 18–28.

SOUTH HALL, MUSEUM OF MODELS, PATENT OFFICE.

FIGURE 9.6 *United States Patent Office Building* (SMITHSONIAN IMAGE 2008-4988).

models in a new building (*Fig. 9.6*). There was even sufficient interest in the subject to justify publication of a catalogue of the location of various models.[23]

According to this 1855 catalogue, three cases in the new hall were devoted to "Mathematical, Philosophical and Optical Instruments, including Clocks, Chronometers, &c." The catalogue lists only the general topic of an invention and the number of models of such inventions, not the inventor. In some cases, it is not difficult to match things up with surviving patent models in the Smithsonian collections. It seems likely, for example, that the "machine for adding numbers" shown at the Patent Office is the device so described in U.S. Patent 3574 taken out by Jehu Hatfield in 1854, as this is the only such object patented in the U.S. between 1836 and 1855. Similarly, it is probable that the "rules for calculating interest" are an invention of one S.S. Young. With slide rules, the case is more ambiguous. The Smithsonian collections include the patent

23 Alfred Hunter, *A Catalogue Showing the Location of Every Model of Patented Inventions in the New Hall of the Patent Office with the Class to Which It Belongs . . .*, Washington, D.C., 1855, pp. 42–43.

model for a circular slide rule invented by John W. Nystrom of Philadelphia and patented as an "improvement in calculating machines" (*Pl. 10*). According to Hunter's catalogue, three models of improvements in machines for calculating were on display. Patent records indicate that five objects meeting this description had been patented. Thus one can say that it is probable but by no means certain that Nystrom's circular slide rule went on display at the U.S. Patent Office in the mid-nineteenth century.[24]

As the number of patents granted burgeoned, the country expanded, and inventors gradually came to specialize, maintaining a complete collection of patent models became unworkable. In 1871, the Patent Office began printing full copies of patents and patent drawings. These were distributed to the inventor and to various depository libraries and sold to the general public. The scheme proved successful. After an 1877 fire destroyed at least 76,000 patent models, the Patent Office dropped the model requirement altogether. By the early twentieth century, it was ready to part with the remaining models. Some were transferred to the Smithsonian, others were distributed to schools and other museums, and some were sold at public auction.[25] The model of Nystrom's invention came to the Smithsonian from the Patent Office in 1908.

Some of the patent models received from the Patent Office were put on display at the Smithsonian in 1910. Whether Nystrom's slide rule was one of them we do not know. Smithsonian records as to precisely what has been on display are even more fragmentary than those of the Patent Office. Collecting and exhibiting slide rules otherwise also began slowly, primarily as a byproduct of other collecting programs.

More generally, it was only in the mid-twentieth century, when a separate museum of history and technology was planned, that the Smithsonian set out to systematically acquire a few slide rules. Some of these were exhibited as part of the mathematics hall from the mid-1960s. Others appeared in various temporary exhibits. A few slide rules, including the Fuller Time Telegraph, made their way into a section on data processing in a more permanent 1989 exhibit. As one might guess, they did not dominate the show. For many years, a small slide rule appeared as part of a reconstruction of Enrico Fermi's Chicago Pile in an exhibit on science in American life. Demonstration slide rules have also appeared in more recent exhibits at the National Air and Space Museum as

24 For a manual for Nystrom's circular slide rule, see George Thorsted, *A Manual for Nystrom's Calculator*, Philadelphia, 1854.

25 *American Enterprise: Nineteenth-century Patent Models: An Exhibition Organized by Cooper-Hewitt Museum ...*, New York, 1984.

well as the National Museum of American History. Additionally, NMAH staff continue to collect them in small numbers.

Slide Rules Online

Collecting slide rules is not a new activity. Firms that produced them, such as Keuffel & Esser, saved examples of their own products and those of rival companies. A few mathematicians and historians interested in the history of the instrument created displays, particularly at the time of the tercentenary of Napier's discovery of logarithms in 1914. After the abrupt disappearance of slide rules from retail shelves in the 1970s, a much wider group of nostalgic users began building collections of the instrument. Individual collectors gradually found each other and formed organizations, including the Oughtred Society in the United States (1991) and 'circles' in the Netherlands (1991), the United Kingdom (mid-1990s), and Germany (ca. 1999).[26] Members exchanged and displayed rules at their meetings, but in addition, they started sharing their holdings on the web, an interface that of course had also made it much easier to advertise and purchase slide rules through auction sites such as eBay. Slide rules are now displayed online on websites that range from images of a few rules with minimal captions to extensive virtual exhibits with detailed background information on instruments or makers.[27] Generally, websites allow their creators to share their enthusiasm for slide rules, but some additionally keep the memory of the object alive by offering training to readers through animations of slide rules in operation. Notable aids for researchers, in addition to the International Slide Rule Museum mentioned above, include David M. Riches's pages depicting a variety of mathematical instruments, the maker archives hosted by the Sphere Research Corporation of Canada, and Clark McCoy's comprehensive gathering of scans from Keuffel & Esser catalogues.[28]

26 The Oughtred Society, www.oughtred.org; Dutch Circle of Slide Rule Collectors, www.rekeninstrumenten.nl/Engels/english.htm; United Kingdom Slide Rule Circle, www.uksrc.org.uk; Deutschsprachige Rechenschieber-Sammler, www.rechenschieber.org (all accessed 11 March 2014).

27 Greg Scott, a citizen of Australia, compiled a list of collectors on http://sliderule.ozmanor.com/srlcol.html (accessed 11 March 2014). Although the list has not been updated since 2004, many of the links remain active if not necessarily updated by their creators.

28 David M. Riches, *Mathematical Instruments: A Private Collection*, www.mathsinstruments.me.uk/index.html; Walter Shawlee II *et al.*, The Slide Rule Universe Archives, www.sphere.bc.ca/test/2archives.html; Clark McCoy (ed.), *Keuffel & Esser*, www.mccoys-kecatalogs.com (all sites accessed 11 March 2014).

For the most part, these websites predated efforts by museums to digitize their collections. Since these sites also are often not the work of professional web designers, they offer both positive and negative lessons about arranging images and determining the content of text. Further, they focus specifically on slide rules, while museums generally display the instrument among a great mass of other objects. This means that the average person who decides to search for slide rules on a search engine such as Google will be taken directly to the collectors' sites, but to find individual catalogue records prepared by curators and trained volunteers he or she first must identify a particular museum that may own slide rules and then search within that institution's online collections.[29] More recently, museums have begun organizing related records into mini-exhibits devoted to a particular instrument.[30] It is likely that forms of online displays will continue to evolve with changes in technology as well as political pressures to increase the accessibility of publicly-funded institutions and budgetary pressures to accomplish this increased outreach with less funding.

In this time of rapidly changing digital displays, looking to the past offers useful insights. The way visitors to a web site approach digital images will be, if possible, even more random than nineteenth century arrangements at a world's fair or at the Patent Office. How the images will be used is perhaps as unpredictable as the way people once wore slide rules. We might guess, however, that slide rules are likely to be displayed in the digital world among a great mass of objects, much as they have been in many museum exhibits.

29 Some of the larger museum collections online at present include: *MIT Museum Collections*, http://webmuseum.mit.edu/, with 545 rules by Keuffel & Esser and other makers; Smithsonian Institution, *Collections Search Center*, http://collections.si.edu, with 209 slide rules and more than 34 instruction manuals and other documentation; Harvard University, Department of the History of Science, *Waywiser*, http://dssmhi1.fas.harvard.edu/emuseumdev/code/eMuseum.asp?lang=EN, with 218 slide rules and links to histories of manufacturers; and Museum of the History of Science, Oxford, www.mhs.ox.ac.uk/collections, with 107 slide rules and related documentation (all sites accessed 25 October 2012).

30 For example, see Smithsonian's National Museum of American History, *Object Groups*, http://americanhistory.si.edu/collections/object-groups (accessed 11 March 2014).

CHAPTER 10

"Exceedingly Ridiculous": Telescopes on Display on the Seventeenth-Century Stage

Ingrid Jendrzejewski

Introduction

In 1610, Galileo Galilei published *Sidereus Nuncius*, the first work in which he presented his telescope-based observations. Within twenty-four hours of publication, Sir Henry Wotton, the English ambassador to Venice, wrote to the English King James I that Galileo "hath first overthrown all former astronomy" and that "the author runneth a fortune to be either exceedingly famous or exceedingly ridiculous".[1] Among those who found much that was exceedingly ridiculous in Galileo's work were a number of playwrights who wrote for the English stage. This paper seeks to explore some of the ways seventeenth-century English theatre responded to advancements in astronomy and, in particular, to Galileo and the telescope.

Telescopes on Stage

An early mention of a telescope on stage can be found in Ben Jonson's recently rediscovered *Entertainment at Britain's Burse*. This masque pre-dates the publication of *Sidereus Nuncius* by a year, being first performed on April 11, 1609. The entertainment was commissioned by the Earl of Salisbury on the occasion of the opening of Britain's Bourse, and it was performed in the market itself.[2] In it, Jonson celebrates the luxury goods and novelties available for purchase there. Among the wonderful items mentioned are a number of scientific instruments, including a so-called 'perspective':

1 Marie Boas Hall, *The Scientific Renaissance: 1450–1630*, Mineola, 1994, p. 320.

2 James Knowles (ed.), "The Entertainment at Britain's Burse", in Richard Dutton (ed.), *Epicene, Or, The Silent Woman: The Revels Plays*, Manchester, 2003, pp. 279–298; 279–280.

 | DOI 10.1163/9789004264403_011

> Master: But here's my jewel: my perspective. I will read you with
> this glass the distinction of any man's clothes ten, nay
> twenty mile off, the colour of his horse, cut or long tail,
> the form of his beard, the lines of his face.
> Boy: If it be toward you, sir!
> Master: If it be but half, I care not! Nay, I will tell by the
> moving of his lips what he speaks and in what language.
> If the sun shine anything strong, I will stand you in
> Covent Garden familiarly and decipher at Highgate the
> subtlest carracts you can make as easily as here [. .].[3]

At the time, the word 'perspective' could refer to any optical device that one looked through—anything from a magnifying glass to a monocle to a telescope.[4] Editor James Knowles annotates Jonson's use of the word 'perspective' as such, stating that the instrument is "Possibly a telescope, although in 1609 telescopes were great rarities, mainly sourced in Holland from 1608 onwards. They were extremely expensive, to such an extent that the Council of Venice hired Galileo to fashion one rather than pay the high price demanded by travelling salesmen".[5]

A reasonable case can be made for the telescope reading. First, the perspective is described as being very rare—the 'jewel' of the Master's collection. Second, several other optical devices can be ruled out as they are mentioned earlier in the text and would not be capable of such great magnification.[6] If the Master is not exaggerating, it is capable of much greater magnification than single-lens devices. Third, word of the telescope spread quickly across Europe after its invention, with physical telescopes being available in Paris by 1609. As Henry King writes:

> While in 1608 the French ambassador at The Hague was negotiating for a telescope to send to Henry IV, early the next year telescopes were on sale in Paris. At the Frankfurt fair in the autumn of 1609, an itinerant Belgian showed a telescope which magnified 'several times' to a friend of Simon Marius, the astronomer. In May, *Dutch trunks*, *perspectives* or *cylinders*

3 Ibid., p. 291.

4 "Perspective, n.", Def. 2a, *Oxford English Dictionary*, Online Ed., www.oed.com (accessed June 2013).

5 Knowles, op. cit. (n. 2), p. 291.

6 Ibid., pp. 290–291.

> as they were then called, appeared in Milan, a little later in Venice and Padua and, by the end of the year, they were being made in London.[7]

This may be the only telescope reference on the English stage that pre-dated the *Sidereus Nuncius.* Unsurprisingly, the perspective is held up as a rare and novel object, and Jonson describes its power of magnification in such a way as to inspire awe and wonder in his audience—an audience who is not necessarily expected to have heard of such an instrument. Within two years of the publication of *Sidereus Nuncius*, however, word of Galileo and his telescope appears to have travelled quickly enough that playwrights feel confident referencing them. Possibly the earliest mention of a telescope in a stage play to appear after Galileo published is this casual reference found in John Webster's 1612 production, *The Duchess of Malfi*:

> Cardinal: We had need goe borrow that fantastique glasse
> Inuented by *Galileo* the Florentine,
> To view an other spacious world i'th' Moone,
> And looke to finde a constant woman there.[8]

The play is set in the early sixteenth century so the telescope reference is anachronistic, but the meaning is clear: to the cardinal, the idea of a faithful woman is so ridiculous that the only hope of finding one is to look to the Moon. Although the telescope is still described as 'fantastique', these passing references clearly suggest that Webster expects his audiences to have heard of Galileo and his observations.

Telescopes feature in more substantial ways in two academic plays that were first performed within the same decade as *Sidereus Nuncius.* The first is Thomas Tomkis's *Albumazar*, an academic play based on Giovan Battista della Porta's 1606 play, *Lo Astrologo.*[9] There is some dispute about the date *Albumazar* was written, but the first production was commissioned by Trinity College, Cambridge to entertain King James I during a March 1614 visit.[10]

---

7 Henry C. King, *The History of the Telescope*, New York, 1955, p. 33.

8 John Webster, *The Tragedy of the Dvtchesse of Malfy, As it was Presented priuatly, at the Black-Friers; and pvbliquely at the Globe, By the Kings Maiesties Servants*, London, Printed by Nicholas Okes, for Iohn Waterson, and are to be sold at the signe of the Crowne, in Paules Church-yard, 1623, Act II, Scene 4; facsimile reprint edition, Menston, 1968, n. p.

9 Hugh Dick (ed.), *Albumazar: A Comedy*, London, 1944, p. 12.

10 Ibid., pp. 53–54.

The title character of *Albumazar* is a charlatan who is one of the first of the 'cheating astrologer' characters that would become popular on stage in the seventeenth century.[11] The telescope is employed for purposes of deceit. In one scene, Albumazar's sidekick Ronca attempts to dupe the elderly Pandolfo by demonstrating his so-called 'perspicill'; when he tells Pandolfo he has the ability to "see the heauens", the following dialogue ensues:

> Pandolfo: What's this I pray you?
> Ronca: An engine to catch starres,
> A mace t'arrest such Planets as haue lurk't
> Foure thousand yeares vnder protection
> Of *Iupiter* and *Sol*.
> Pandolfo: Pray you speake English.
> Ronca: Sir, 'tis a Perspicill, th'best vnder heauen:
> With this I'le read a leafe of that small *Iliade*
> That in a Wall-nut shell was desk't, as plainly,
> Twelue long miles off, as you see *Pauls* from *Highgate*.[12]

Ronca carries on, describing all manner of things that can be seen through the telescope and then convinces Pandolfo to try out another one of Albumazar's instruments, the otacousticon, which is meant to magnify sound in much the way a telescope magnifies things that one can see. Quite appropriately, the otacousticon is shaped like a gigantic pair of ass's ears.

In his 1943 paper, *The Telescope and the Comic Imagination*, Dick makes the point that in *Albumazar*, Tomkis is very directly engaging with *Sidereus Nuncius*. Not only does Tomkis invoke Galileo's name, he uses the term 'perspicill' to describe the telescope where Galileo uses 'perspicillum' in *Sidereus Nuncius*.[13] Galileo does not describe his telescope in great technical detail; he introduces the instrument briefly and then quickly transitions to his descriptions of the amazing things he has been able to observe with it; similarly, Ronca

11 Ibid., p. 47.

12 Thomas Tomkis, *Albumazar. A Comedy presented before the Kings Maiestie at Cambridge, the ninth of March 1614. By the Gentlement of Trinitie Colledge*, London, Printed by Nicholas Okes for Walter Burre, and are to be sold at his Shop, in Pauls Church-yard, 1615, Act I, Scene 3, n. p.

13 Dick, "The Telescope and the Comic Imagination", *Modern Language Notes* 58 (1943), pp. 544–548: 545. I also credit Dick for introducing me to a shortened version of the Wotton quote which frames this paper.

spends multiple lines describing what can be seen through his perspicill.[14] Both Galileo and Albumazar discover new planets with their telescopes,[15] and, further, where Galileo chooses to name the moons of Jupiter the Medicean Stars in honour of his patron, Albumazar suggests naming his newly-discovered stars after the man he is trying to con:[16]

> Albumazar: *Ronca*, the bunch of planets new found out
> Hanging at th'end of my best Perspicill,
> Send them to *Galilaeo* at *Padua;*
> Let him bestow them where hee please. But th' starres
> Lately discouered 'twixt the hornes of *Aires*,
> Are as a present for *Pandolfo's* marriage,
> And henceforth stil'd *Sidera Pandolfaea*.[17]

Tomkis was not the only academic who referenced Galileo and put telescopes in the hands of astrologers. On the 23rd of February in 1617, Barten Holyday's *Technogamia: or The Marriages of the Arts* was first performed at Christ Church, Oxford and, despite it being met with "no great applause", the show was revived in 1621 and performed before King James I at Woodstock, again to a less than favourable reception.[18]

Holyday's allegory dramatises the arguments and intrigues between the arts and sciences, and shows the triumph of Physica and her daughter Astronomia over Magus and his wife Astrologia. The plot is rather episodic, with various disciplines attempting to form inappropriate alliances, such as when Poeta seeks the hand of the highly-sought-after Astronomia. Order is restored at the end, however; Magus and Astrologia are banished from the land and Astronomia pairs up with Geographus.

Interestingly, it is Magus, not Astronomia who talks of telescopes:

> Magus: [...] But there's one *Galilaeus* an exquisite Mathematicean, an *Italian*: whom I came very lately acquainted with, by admirable lucke; and he has promis'd to helpe me to a glasse, by which I shall see all

14 Ibid., p. 545.

15 Ibid., pp. 546.

16 Ibid., pp. 546–547.

17 Tomkis, *Albumazar*, cit. (n. 12), Act I, Scene 5, n. p.

18 Anthony Wood, *Athenae Oxonienses, The Second Edition,* London, Printed for R. Knaplock, D. Midwinter, and J. Tonson, 1721, 3 vols.: v. 2, n. p., cols. 259–260.

> things as perfectly represented in *Astronomia's* house, as if I were there: till which time I must take the paines to haue it by relation [...].[19]

Instead of being used to observe the heavens, the telescope is co-opted by Magus to aid in his plan to spy on Astronomia.

It was not only the academic plays that engaged with Galileo's work; several playwrights writing for larger audiences were inspired by the idea that the Moon, like the Earth, has mountains and valleys. One of the first of these was Ben Jonson in his 1620 masque, *News from a New World Discovered in the Moon*, in which the idea of an inhabited moon was used to poke fun at the newly-developing news industry. The masque begins with two heralds announcing to a printer, a chronicler and a factor that they have a potential news story that has come "from the Moone". There is some initial confusion when the Factor asks how news was obtained as the Printer immediately assumes that news must have come via a 'trunk' (telescope):

> Printer: Oh, by a trunk! I know it, a thing no bigger than a flute case. A neighbour of mine, a spectacle-maker, has drawn the moon through it at the bore of a whistle, and made it as great as a drumhead twenty times, and brought it within the length of this room to me, I know not how often.[20]

These, however, are modern times; as the Chronicler reminds us, telescopes are old news now:

> Chronicler: Tut, that's no news; your perplexive glasses[21] are common.

The heralds make it clear that the method of transport is nothing as confusing as a "Mathematician's perspicil",[22] but that the news has come via "the neat and clean power of poetry", "The mistress of all discovery".[23] Galileo may have

19 Barten Holyday, *Technogamia: or The Marriages of the Arts. A Comedy*, London, Printed by Iohn Haviland for Richard Meighen, and are to be sold at his shop next the middle Temple gate, and in Saint Dunstans Church-yard in Fleetstreet, 1630, Act IV, Scene 4, n. p.

20 Ben Jonson, "News from the New World Discovered in the Moon: A masque, as it was presented at court before King James, 1620", in Stephen Orgel (ed.), *Ben Jonson: The Complete Masques*, London, 1969, p. 295.

21 Here, the Chronicler humorously calls the telescope a 'perplexive' instead of a 'perspective' glass.

22 Ibid.

23 Ibid., p. 296.

discovered mountains and craters on the Moon, but through poetry the heralds have discovered that the Moon is home to all manner of terrestrial and fantastic features including everything from landscape to lovers and from law to lapdogs—or moon-calves, as they are supposedly called on the Moon. For much of the rest of the masque, the heralds describe the features of the world in the Moon, and later, the audience is treated to a dance performed by Volatees, a race of feathered Moon people.

Jonson makes yet another telescope reference five years later in *The Staple of News*, a much longer satire of the news industry. The play begins with a character called Pennyboy inviting the audience to:

> Pennyboy Junior: [...] looke on me, and with all thine eyes,
> Male, Female, yea, *Hermaphroditicke* eyes,
> And those bring all your helpes, and perspicills,
> To see me at best aduantage, and augment
> My forme as I come forth, for I doe feele
> I will be one, worth looking after, shortly.[24]

By 1625, only sixteen years after Jonson's Master went to great lengths to describe the things one could see through the 'jewell' of the Master's collection of scientific instruments, Jonson is so confident that his audience is familiar with telescopes that he begins his play with a casual reference.

English theatres were closed in 1649 during the Interregnum, so much less is written for the stage until the theatres reopened in 1660—the same year that the Royal Society was founded. Although most characters who carried telescopes on the seventeenth-century stage were not meant to be taken seriously, there were a few exceptions, such as can be found in two plays that premièred in the 1660s.

In 1662, Margaret Cavendish wrote *Love's Adventures*, one of the few plays in which the man involved in studious astronomical pursuits—Sir Peaceable Studious—is not presented as a comedic figure. Instead, Cavendish invites her audience to laugh at Sir Peaceable's wife, Lady Ignorant, whose lack of appreciation of her husband's intellectual activities makes her appear unlearned and ridiculous:

---

24 Jonson, "The Staple of Nevves. A Comedie Acted in the Yeare, 1625 by His Maiesties Servants", in *The workes of Benjamin Jonson. Vols. 2–3*, London, Printed by I.B. for Robert Allot, and are to be sold at the signe of the Beare in Pauls Churchyard, 1631 [i.e. 1641], Act I, Scene 1, n. p.

Lady Ignorant: Lord Husband! I can never have your company, for you are at all times writing, or reading, or turning your Globes, or peaking thorough your Prospective Glasse, or repeating Verses, or speaking Speeches to your self.

Sir Peaceable Studious: Why wife, you may have my company at any time, Nay, never to be from me if you please, for I am alwaies at home.[25]

Despite his preference for a quiet, solitary life, the portrayal of Sir Peaceable Studious is quite positive, reflecting the author's interest in experimentation and research. As Eleanor Mattes describes, Margaret Cavendish's: "*Observations in Experimental Philosophy*, published in 1663, shows genuine familiarity with the telescope and the microscope. And respect for her husband's position brought a favorable response to her request to attend a meeting of the Royal Society, at which she was shown 'severall Experiments'".[26]

Another not altogether negative depiction of scholars who associate with telescopes is found in St Serfe's only play, *Tarugo's Wiles: or, the Coffee-House* which premièred in 1667 and "had a great popularity in the first years of the Restoration".[27] The entire play is set in a coffee-house filled with "a mixture of all kind of people",[28] and in Act III the customers listen in on an extended conversation between two scholars who are positioned "at the Table where the Globes were, the one is instructing the other in the Celestial-Globe":[29]

1st Scholar: This same gravelike Matron is *Cassiopeia, Jupiter's* Dry-Nurse, and Governess to *Juno*'s Milk-maids.

2nd Scholar: Is this she who with her Troop of Country-wenches goes every *May-day* a milking Cowes in *Via-Lactaea*.

---

25 Margaret Cavendish, "Loves Adventures", in *Playes Written by the Thrice Nobel, Illustrious and Excellent Princess, the Lady Marchioness of Newcastle*, London: Printed by A. Warren, for John Martyn, James Allestry, and Tho. Dicas, at the Bell in Saint Pauls Church Yard, 1662, pp. 13–14.

26 Eleanor Mattes, "The 'Female Virtuoso' in Early Eighteenth Century Drama, *Women & Literature* 3:2 (1975), 3–9: p. 4. Mattes cites E.S. de Beer (ed.), *The Diary of John Evelyn*, Oxford, 1955, III, p. 483.

27 Allardyce Nicholl, *A History of English Drama 1660–1900: Volume I Restoration Drama 1660–1700*, London, 1967, p. 220.

28 Thomas St Serfe, *Tarugo's Wiles: or, the Coffee House. A Comedy. As it was Acted at his Highness's, the Duke of York's Theatre*, London, Printed for Henry Herringman, at the Sign of the Anchor, on the Lower-walk of the New-Exchange, 1668, Dramatis Personae, n. p.

29 Ibid., p. 21.

> 1st Scholar: The same: and there they make provisions of Cheese-cakes and Creame for the Mathematical Feast in *Copernicus*-Hall; but when I come to demonstrate with *Galileus's* Tube, I'le discover to you all the black-patches on her Face, and whether her Whisk be right set off, and how many tire of Gimp-lace circles the skirts of her Petticoat.[30]

The scholars are incidental to the plot and not even listed in the cast of characters in the 1668 printing, but they serve to represent the scholarly patrons in the seventeenth-century coffee-house scene. Their scholarly credentials are confirmed with a bit of Latin, a bit of philosophy, knowledge of the constellations, and, of course, the promise of a telescope demonstration. At the time, coffee-houses often sponsored cheap lectures and demonstrations, making this an important scene in establishing the play's setting.

The more positive portrayals of telescope users in *The Basset-Table* and *Tarugo's Wiles* are rare; as the century progressed, the Royal Society's 'new science' provided Restoration playwrights with great fodder for comedy. In the latter part of the century, many ridiculous practitioners of experimental science could be found on the English stage.

Thomas Shadwell's 1676 production, *The Virtuoso*, for example, directly references—and ridicules—many of the experiments described in the Royal Society's *Philosophical Transactions*, Thomas Sprat's *History of the Royal Society* (1667) and Robert Hooke's *Micrographia* (1665).[31] Astronomical observation is but one of the branches of science that come under fire; as Claude Lloyd states in 1929,

> For the substance of his satire, Shadwell began by choosing those phases of science, or pseudo-science, which seemed most ridiculous when subjected to the common sense of the time. Such, in science itself, were the ideas of the moon as a planet with a physiography of its own [...] Astrology and the occult sciences were as amusing for being behind as was experimental science for being ahead of the times.[32]

Sir Nicholas Gimcrack is the virtuoso to whom the title refers and, in 1916, Carson S. Duncan identifies Gimcrack as the first of a new type of scientist

---

30 Ibid., pp. 21–22.

31 Claude Lloyd, "Shadwell and the Virtuosi", *PMLA* 44 (1929), pp. 472–494: 475. In this paper, Lloyd presents a detailed discussion of the origins of many of the experiments in *The Virtuoso*.

32 Ibid., p. 491.

to appear in stage plays, noting that "the name of Sir Nicholas Gimcrack soon became attached to all manner of scientific apparatus, and was synonymous for years with scientific crankism".[33]

Gimcrack's observation of the Moon is just one of the scientific pursuits that make him look ridiculous, as in this exchange:

> Sir Nicholas Gimcrack: But I wonder Sir *Formal* is not return'd; I sent him to fix my Tellescopes for surveying the Moon.
>
> Longvil: Do you believe the Moon is an Earth, as you told us?
>
> Sir Nicholas Gimcrack: Believe it! I know it; I shall shortly publish a Book of Geography for it. Why, 'tis as big as our Earth; I can see all the Mountainous parts, and Vallies, and Seas, and Lakes in it; nay, the larger sort of Animals, as Elephants and Camels; but publick Buildings and Ships very easily. I have seen several Battels fought there. They have great Guns, and have the use of Gun-powder. At Land they fight with Elephants and Castles. I have seen 'em—.[34]

Like Jonson's newsmen, Gimcrack goes on to describe many details of the world in the Moon, despite the remark from a gentleman "of wit and sense"[35] that "No Phanatick that has lost his Wits in Revelation, is so mad as this Fool".[36]

Here, Shadwell uses the word 'tellescope' to describe Gimcrack's instruments. Although John Bainbridge uses the word 'telescopium' as early as 1619 in a description of his observations of a comet,[37] this may be one of the earliest uses of the word 'telescope' on the English stage.

There is some debate over whether *The Virtuoso* was intended as a critique of the Royal Society or whether "the real scientists, men like Hooke and Boyle, Ray and Newton, were then as always beyond the reach of ridicule"[38] and, thus, "in their grotesque absurdities, both Sir Nicholas and Sir Formal are, then,

33 Carson S. Duncan, "The Scientist as a Comic Type", *Modern Philology* 14 (1916), pp. 281–291: 284.

34 Thomas Shadwell, *The Virtuoso. A Comedy, Acted at the Duke's Theatre*, London, Printed by T.N. for Henry Herringman, at the Anchor in the Lower Walk of the New Exchange, 1676, pp. 70–71.

35 Ibid., Drammatis Personae, n. p.

36 Ibid., p. 71.

37 John Bainbridge, *An Astronomicall Description of the late Comet from the 18. of Nouemb. 1618 to the 16. of December following*, London, 1619; Facsimile reprint, Norwood (New Jersey), 1975, p. 19.

38 Walter E. Houghton Jr., "The English Virtuoso in the Seventeenth Century: Part II", *Journal of the History of Ideas* 3 (1942), pp. 190–219: 211.

aberrations of their respective professions".[39] Either way, Gimcrack is the first of a long line of comic virtuoso characters that take to stage in the seventeenth and eighteenth centuries. As Duncan puts it,

> This is the type of the new scientist as comedy first found him [...]. The man of the new science as the playwriters misinterpreted him, was a fool, because he was engaged in the vain pursuit of useless knowledge, a pedant, because he was a mere pretender to learning, and a generally despicable character, because he was wholly absorbed in a low, vulgar interest outside the social realm of London society folk.[40]

A telescope-toting virtuoso also features in Aphra Behn's second-most-performed play,[41] her 1687 farce, *Emperor of the Moon*, a play "derived directly from *Arlequin Empereur dans le Monde de la Lune* (1684)" and introducing "the typical figures of the *commedia dell'arte*—Harlequin, Scaramouch and Dr Baliardo".[42] The plot revolves around Dr Baliardo, a man so obsessed with the Moon that he neglects his daughter and cannot view the world around him with the objectivity required of a competent natural philosopher.

Physical humour abounds in *The Emperor of the Moon*, and Behn invites us to laugh at Baliardo and his large instruments, with phallic overtones very much intended. For example, when Baliardo is first introduced, Behn writes in the stage directions: "Enter Doctor, with all manner of Mathematical Instruments, hanging at his Girdle; Scaramouch bearing a telescope twenty (or more) Foot long".[43] In another instance, Baliardo is in a hurry to view what he thinks are private activities in the royal chamber of a race of Moon people; as his servant wrestles with the large telescope, trying to position it properly, a frantic Baliardo cries out, "Mount, mount the telescope".[44]

The other characters, at best, are frustrated by Baliardo's instruments; "Run, Run *Scaramouch*," says the governess, "my Masters Conjuring for you like Mad below, he calls up all his little Divels with horrid Names, his *Microscope*, his *Horoscope*, his *Telescope*, and all his *Scopes*".[45]

39 Joseph M. Gilde, "Shadwell and the Royal Society: Satire in The Virtuoso", *Studies in English Literature, 1500–1900* 10 (1970), pp. 469–490: 490.

40 Duncan, "The Scientist", cit. (n. 33), pp. 287–288.

41 Jane Spencer (ed.), *The Rover and Other Plays*, Oxford, 1995, p. xxi.

42 Nicholl, *English Drama*, cit. (n. 27), p. 262.

43 Aphra Behn, *The Emperor of the Moon: A Farce. As it is Acted by Their Majesties Servants, at the Queens Theatre*, London, Printed by R. Holt for Joseph Knight, and Francis Saunders, at the Blew-Anchor in the lower Walk of the New Exchange, 1687, p. 7.

44 Ibid.

45 Ibid., p. 5.

At worst, other characters use Baliardo's telescopes to deceive him, such as in this episode in which one of the romantic interests fools Baliardo into thinking that a nymph can be seen through the telescope. Baliardo is told, "If you be throwly purg'd from Vice, the opticles of your sight will be so illuminated, that glancing through this *Telescope*, you may behold one of these lovely Creatures, that people the vast Region of the Air", and then, whilst Baliardo looks through the telescope, "Charmante goes to the Door to Scaramouch, who waited on purpose without, and takes a Glass with a Picture of a Nymph on it, and a light behind it; that as he brings it, it shows to the Audience. Goes to the end of the Telescope". After Baliardo is convinced to pray to the Spirit of the East, "Charmante puts the Glass into the Mouth of the Telescope",[46] and Baliardo is in raptures of delight, thinking he has seen a nymph. Don Charmante follows this up by placing another glass plate onto the telescope, this time with a picture of the supposed Emperor of the Moon and the ruse works a second time.[47]

This is only the beginning of the trickery in Behn's farce. Baliardo's daughter and niece have lovers whom they want to marry. The two couples, along with the wily servants Scaramouch and Harlequin, decide to stage an elaborate hoax to fool Baliardo into thinking that he is on the Moon, in the court of its emperor. Using props, costumes and a large empty building on Baliardo's estate, they build a fanciful environment that they pass off to Dr Baliardo as the world in the Moon, telling him that the Emperor of the Moon and the Prince of Thunderland saw his daughter and niece through their telescopes and fell desperately in love. The final act takes place in what Dr Baliardo believes to be the court of the Emperor. The spectacle is immense; no effort is spared to overwhelm Dr Baliardo with the majesty of the supposed Moon-world. The doctor is even introduced to Galileo and Kepler figures as this stage direction indicates:

> *The Scene in the Front draws off, and shews the Hill of Parnassus; a noble large Walk of Trees leading to it, with eight or ten Negroes upon Pedestals, rang'd on each side of the Walks. Next* Keplair *and* Gallileus *descend on each side, opposite to each other, in Chariots, with Perspectives in their Hands, as viewing the Machine of the Zodiack. Soft Musick plays still.*[48]

Galileo and Kepler salute Baliardo on behalf of the Emperor and Prince, then the entire zodiac is invited to descend from the heavens and sing and dance

46 Ibid., pp. 9–10.

47 Ibid., p. 11.

48 Ibid., p. 59.

for the doctor in one of the many spectacles performed before him. The doctor is beside himself with awe throughout, never questioning what he sees before his eyes. When the conspirators finally tell Baliardo that he is not on the Moon, he realises the folly of his ways and renounces his scientific enquiry, exclaiming:

> Baliardo: Burn all my Books, and let my Study Blaze,
> Burn all to Ashes, and be sure the Wind
> Scatter the vile Contagious Monstrous Leys.[49]

In *The Emperor of the Moon*, the virtuoso's scientific instruments mark him as a ridiculous man. Instead of helping him to learn more about the universe, Baliardo's telescope is the instrument of his downfall; Baliardo is unable to distinguish what he thinks he is observing from reality.

It is not only male characters who are ridiculed for their scientific pursuits, however; lady virtuosos become "almost a stock character in late seventeenth and early eighteenth century English satire",[50] and, of course, several of these lady virtuosos can be found looking through telescopes.

Some such women appear in Thomas Wright's 1693 comedy, *The Female Virtuoso's*, a comedy largely based on Molière's 1672 production *Les Femmes Savantes*.[51] Catchat ("a stale Virgin, who fancies every Man is in Love with her")[52] talks about her observations of people on the Moon,[53] and, later, Catchat and her niece, Lovewitt ("an Admirer of Platonic Love, yet in Love with Clerimont") make fools of themselves by talking about telescopes in front of Sir Maurice Meanwell ("an honest rich Citizen"):[54]

> Sir Maurice Meanwell: A pretty thing indeed, to see those long Spectacles of yours, set on the Top of my House, for you to peep, and tell how many Hackney Coaches are going in the Moon.
> Lovewitt: Oh the Illiterate Brute! thus to affront a Telescope. (aside)[55]

49 Ibid., p. 67.

50 Mattes, "'Female Virtuoso'", cit. (n. 26), p. 5.

51 Nicholl, *English Drama*, cit. (n. 27), pp. 186 and 188.

52 Thomas Wright, *The Female Virtuoso's. A Comedy: As it is Acted at the Queen's Theatre, By Their Majesties Servants*, London, Printed by J. Wilde, for R. Vincent, in Cliffords-Inn-lane, Fleet-street, 1693, Dramatis Personae, n. p.

53 Ibid., p. 22.

54 Ibid., Dramatis Personae, n. p.

55 Ibid., p. 25.

Not only are Catchat and Lovewitt made foolish by their learning, but they are explicitly told that "all the Study and Philosophy of a Wife, should be to please her Husband, instruct her Children, have a Vigilant Eye over Domestic Affairs, keep a good Order in her Family, and stand as a Living Pattern of Virtue, and Discretion to all about her".[56]

A more sympathetic lady virtuoso features in Susannah Centlivre's 1706 comedy, *The Basset-Table*, where the heroine Valeria, a "Philosophical Lady",[57] is shown to be more interested in the natural sciences than in marriage and social engagements. She dissects animals and insects, is adept at using a microscope, and demonstrates her familiarity with the telescope when she asks a would-be suitor:

> Valeria: I would have ask'd you, Sir, if ever you had the Curiosity to inspect a Mermaid—Or if you are convinc'd there is a World in every Star—We by our Telliscopes, find Seas, Groves and Plains, and all that; but what they are Peopled with, there's the Query.[58]

Although Valeria talks of mermaids and people on the Moon, she is portrayed in a reasonably positive light; unlike other ladies in the play, Valeria does not have to renounce her interests in order to obtain a favourable romantic match.

From the 1670s onward, playwrights routinely begin to expect audiences to be familiar enough with telescopes to understand casual references to the instruments. One of many examples is Thomas D'Urfey's 1679 comedy, *Squire Oldsapp*, in which a man caught in a compromising position plays the innocent by saying:

> Henry: Well; then in brief, we'll give you full satisfaction at home, but no more words of it here—you see we are observ'd; and 'twill be for neither of your Honours to make discovery: In the mean time, believe your Jealousie always gives you the wrong end of the prospect- but hush, more of that anon—[59]

---

56 Ibid.

57 Susannah Cent-Livre, *The Basset-Table. A Comedy. As it is Acted at the Theatre-Royal in Drury Lane, by His Majesty's Servants*, London, 1706, Dramatis Personae, n. p.

58 Ibid., p. 32.

59 Thomas D'Urfey, *Squire Oldsapp: or, the Night Adventurers. A Comedy: As it is Acted at His Royal Highness the Duke's Theatre*, London, Printed for James Magnes and Richard Bentley, in Russel-street in Covent-Garden, near the Piazza, 1679, p. 63.

Here, the cut off word 'prospect-' refers to 'prospective-glass'. Originally, a 'prospective glass' referred not to a scientific instrument but to a magical object that allowed one to see distant or future events.[60] However, during the 1600s, the meanings of 'prospective' and 'perspective' began to merge and the words were often used interchangeably.[61]

Conclusion

This has been only a brief summary of some of the references to Galileo and the telescope occurring in English-language plays during the seventeenth century. Undoubtedly, historians of science, linguists, and specialists in English drama will be able to add to this discussion, and it is hoped that a more complete understanding of how seventeenth-century English theatre responded to Galileo and the telescope can be obtained with further study of archival material, developments in electronic searches and cross-disciplinary collaboration.

60 "Prospective glass, n.", Def. 2, *Oxford English Dictionary*, cit. (n. 4), (accessed June 2013).
61 "Perspective, n.", Def. 2a, Ibid.

CHAPTER 11

Instruments on Movie Sets: A Case Study

Ileana Chinnici, Donatella Randazzo and Fausto Casi

Introduction

Recreating historical locations for cinematographic purposes requires a great deal of care and historical sensibility. It is a painstaking work involving the gathering of a variety of materials, such as costumes, make-up, furniture and draping, in order to make the scene present and real, and to create some convincing background for the characters and the unfolding story, be it a re-enactment of past events or a piece of fictional narrative. In quite a number of movies, antique scientific instruments have played an important role in recreating particular scenes. Even if special replicas are nowadays purpose-made by specialized firms for movie sets, in some cases the use of original instruments still occurs. By "starring" close to famous actors, in some cases these apparatus have gained a high degree of visibility, whereas in other occasions they have just been a mere background presence. In either case, their inclusion in the movie can be regarded as an effective way of contextually displaying them and for the movie-going public an unusual way of experiencing scientific instruments.

Historical instruments may play different "roles" in a movie. They can be presented as tools for contemporary use in historical reconstructions, where they confer authenticity to the set, playing a "historical" role. Or they can maintain their identity as antique objects, thus playing an "aesthetic" role merely furnishing the set of the film. The present paper examines both cases and offers some examples of well-known cinematographic productions which have made use of original scientific instruments on the set; it aims at drawing readers' attention to this unusual subject without pretending to be exhaustive.

The Case of *The Leopard* (1963)

In the early 1960s the art-director Luchino Visconti (1906–1976) was working on the production of *Il Gattopardo* (The Leopard), a cinematographic version of the novel by Giuseppe Tomasi di Lampedusa (1896–1957) dealing with

 | DOI 10.1163/9789004264403_012

the fall of the Bourbon Kingdom in Sicily, as narrated through the events of a Sicilian aristocratic family, headed by the Prince Fabrizio Salina. The leopard, depicted on the coat of arms of the family, was the symbol of the Prince. The author himself, Tomasi di Lampedusa, belonged to an old Sicilian aristocratic family and, in drawing the profile of the protagonist of the novel, he was inspired by the figure of his great-grandfather, the Prince Giulio Fabrizio Tomasi di Lampedusa (1815–1885), a distinct nobleman who was also a renowned amateur astronomer. In 1852 Prince Giulio had built a private observatory in his Villa ai Colli (on the hills) near Palermo and entertained himself by observing astronomical phenomena and publishing the results in local journals. His private observatory was so well equipped with excellent astronomical clocks and telescopes that the astronomer Pietro Tacchini (1838–1905), a pioneer of solar physics in Italy, in 1883 with no hesitation stated that it was "the only Italian private observatory worthy to be mentioned".[1] After the Prince's death, his heirs decided to dismantle the observatory and give up the instruments. The three most significant pieces—a Merz refractor (*Pl. 11*), Worthington altazimuth (*Fig. 11.1*), and Lerebours-Secretan equatorial (*Fig. 11.2*)—were acquired by the Palermo Observatory.

Luchino Visconti was very well-known for his extremely accurate, almost obsessive interior recreation work on the set of his movies. For this reason, he requested from Palermo Observatory the loan of the Prince's original instruments, as well as furniture, books and archival papers to accompany the instruments, in the attempt to create a convincing ambience resembling as close as possible the Leopard's observatory.

From today's perspective, it is surprising that he was allowed to borrow and remove a variety of items from the Observatory's historical collection of instruments and books. No document about this "loan" exists in the Palermo Observatory archives, although some traces still remain, e.g., some brass components of the Merz refractor were darkened to avoid light reflections during shooting and one nineteenth century archival document shows flowers, initials and dates added while filming.[2] The materials belonging to Palermo Observatory can be recognized in the shots of the Prince's private observatory. The sequence starts about 19 minutes from the beginning of the movie and lasts 4 minutes 37 seconds: it is an important and beautiful sequence, featuring

1 See Pietro Tacchini, *Eclissi totali di sole*, Rome, 1888, p. 112. On the equipment of the Observatory, see Ileana Chinnici, "Gli strumenti del Gattopardo", *Giornale di Astronomia* 23 (1997), pp. 24–29.

2 See Ileana Chinnici, Donatella Randazzo, "Old Astronomical Instruments on a Movie Set: The Case of *The Leopard*", *Bulletin of the Scientific Instrument Society* 109 (2011), pp. 9–12.

FIGURE 11.1 *Worthington altazimuth, one of the instruments from Prince of Lampedusa's private observatory*
(*INAF*—OSSERVATORIO ASTRONOMICO DI PALERMO, MUSEO DELLA SPECOLA).

FIGURE 11.2 *Lerebours & Secretan comet seeker, used by the Prince of Lampedusa* (*INAF*—OSSERVATORIO ASTRONOMICO DI PALERMO, MUSEO DELLA SPECOLA).

an intense conversation between the Prince and his friend and assistant Father Pirrone, containing the Prince's famous statement that "everything changes, so that nothing changes"—words which summarize the leading philosophy of *The Leopard* novel. The sequence starts with the Prince climbing up a narrow staircase leading to his observatory. Father Pirrone is sitting at his desk, concentrating on some astronomical calculations.[3] The setting is very accurate: the room is full of relevant books and instruments, mostly belonging to the Palermo Observatory collections (*Fig. 11.3*).[4] After a short animated conversation between the Prince and Father Pirrone, the subject of the conversation is shifted to the political situation of the time, i.e. the declining aristocracy, the rising middle class and the role of the Church. The sequence ends with a bugle call, introducing the next battle scenes. Incidentally, no other sequences located in the Prince's observatory appear in the movie.

Concerning the use of science instruments in cinematography, the case of *The Leopard* is unique, as the materials used for the set belonged to a state astronomical institution. Consequently, its scientific heritage (instruments, books, archives) was protected by legislation concerning the national cultural heritage. It was required that permission from the relevant authorities be granted to the Director of the Observatory for the loan only upon his production of guarantees for the careful use of the material and, obviously, alterations such as the darkening of brass pieces or writing on archival papers should not have been allowed. However, it did happen and two explanations can be given for it. First, awareness of the cultural and historical value of old scientific instruments has emerged only recently within the Italian astronomical community.[5] In the past, the lack of such awareness had led to the loss and/or disposal of many historical instruments, books and documents, artifacts lying around unattended, often out of sight, and decaying with time. Classified as 'old stuff' useless in a 'modern' scientific institution, these materials were often considered even troublesome, owing to the amount of space they were occu-

3 The opening scene is very faithful to the novel, where we read: "[The Prince] went up a long small staircase and flew into the great blue light of the Observatory. Father Pirrone, [...] was sitting engulfed in algebraic formulae"; Giuseppe Tomasi di Lampedusa, *Il Gattopardo,* Feltrinelli, 1957, pp. 44–45 (translation from the Italian by the authors).

4 A complete list is published in Chinnici, Randazzo, op. cit. (n. 2), p. 12.

5 The first recent initiatives to preserve the historical heritage of Italian Astronomical Observatories date back to the 1980s. See Chinnici, "Italian Astronomical Observatories and their Historical Instruments Collections", in Gudrun Wolfschmidt (ed.), *Cultural Heritage of Astronomical Observatories: From Classical Astronomy to Modern Astrophysics*, Hamburg, 2009, pp. 227–231.

FIGURE 11.3 *Photograph taken on the set of the movie* The Leopard *(1962), showing Burt Lancaster as the Prince of Salina in his observatory with Romolo Valli in the role of father Pirrone. Visible on the set are three of Prince of Lampedusa's instruments, together with a variety of items belonging to the Palermo Observatory* (COURTESY OF TITANUS S.P.A.).

pying. Second, this hurried, undocumented and imprudent loan confirms the authority of Visconti, whose magnetic personality usually met no opposition.

The Case of *The Best Offer* (2013)

This film, written and directed by Italian director Giuseppe Tornatore, gives an accurate representation of the world of art auctions and antiques. The protagonist, Virgil Oldman, (played by the actor Geoffrey Rush) is the elderly and esteemed managing director of a prestigious auction house. The movie is centered on his gradual, unaware falling prey to an astute and long-planned plot to steal his priceless collection of female portraits.

Immediately after the first telephone contact with the mysterious female voice, Claire, that will prove fatal to Virgil, the man is portrayed in the course of an auction in which attention is focused on a scientific instrument, a telescope presented as an original made by Galileo Galilei.[6] Put up for auction as the only one extant on the market, it reaches a very high hammer price, thus recalling the title *The Best Offer*.

Away from the set, the object is an old replica of a Galilean telescope, 105 cm long, mounted on an iron tripod with goat-feet. It is made of brass and leather, covered with golden floral decorative engravings (*Fig. 11.4*).[7] In the movie it has a highly 'symbolic' role; it represents an instrument which, in an epoch-making way, has extended our ability to see further, whereas the auctioneer is on the verge of gradually losing his foresight, becoming dramatically 'blind', unable to see what is plotted behind his back.

Another object worth mentioning is a second telescope, visible in the background as part of the furnishings of Virgil's apartment. It is a typically octagonal, wooden Newtonian telescope, probably by Leon Foucault, dating around 1855, 98 cm long, with a mirror of 13 cm diameter (*Fig. 11.5*). Its presence further highlights the role of scientific instruments as ornamental pieces of art. In particular, Virgil's choice to have it in his living room further confirms the gratification he derives from surrounding himself with beauty in the form of static objects, rather than by risking to get involved in dynamic relationships which he is unable to manage and that will prove to be devastating for him.

A third scientific instrument, although neither particularly decorative nor stunningly beautiful, plays a very important role both for its strong symbolic meaning and its 'active' role. It is a late nineteenth-century wax cylinder phonograph, or 'parlograph', purposely restored and put in working condition for the set (*Pl. 12*). It is visible at least in two sequences. The first occurs in the restoration workshop owned by Robert (actor Jim Sturgess), a young craftsman who helps Virgil to assemble an eighteenth-century mechanical android from pieces of rusty gears found in Claire's villa. Whilst engaged in a conversation

6 It is perhaps worth remembering that the only two original Galileo telescopes known today are kept at the Museo Galileo, Florence; see Albert Van Helden, *Istituto e Museo di Storia della Scienza: Catalogue ef Early Telescopes,* Florence, 1999, pp. 30–33, nos. 1 and 2.

7 See Fausto Casi (ed.), *Sotto il segno di Galileo: Luoghi della scienza in Toscana*, Arezzo, 2009, pp. 24–25.

FIGURE 11.4 *Replica of a Galilean telescope with its stand, used for the set in the auction sequence in* The Best Offer (FAUSTO CASI COLLECTION, AREZZO).

FIGURE 11.5 *Octagonal wooden Newtonian telescope, probably by Foucault (ca. 1855), visible on the set in the auctioneer's apartment* (FAUSTO CASI COLLECTION, AREZZO).

about Claire, Virgil does not pay much attention to the fact that Robert is fiddling with the phonograph trying to make it record sound. The reproduction of Virgil's voice confirms that he has succeeded in making the instrument work. The phonograph is shown again toward the end of the film, when it gives 'voice' to the mechanical android, fully reconstructed, repeating over and over a message recorded by Robert, which provides the key to understanding the whole movie plot. The instrument thus assumes a strong symbolic role; it marks the completion of the gradual re-assembling of the automaton, symbolizing the gradual developing of the complex deceit enveloping the auctioneer.

As the Galilean telescope had marked the beginning of the unfolding plan, the phonograph signals its final act. Indeed, both sight and voice are crucial elements in the movie. In the initial part, Virgil cannot see Claire but her voice renders him gradually blinded by his own unexplored emotions, while in the final part, Robert's voice recorded by the phonograph reveals the accomplishment of the awful cheat, thus opening Virgil's eyes. Therefore, the two scientific instruments, extending the sensory faculties, deliberately contrast with the metaphoric blindness and deafness induced by bursting emotions: they appear to play the role of reason opposed to the irrational.

The set of Robert's restoration workshop was carefully designed by Tornatore. A great number of instruments and devices visible in the laboratory scenes come from Fausto Casi collection, Arezzo, as well as from the Museo dei Mezzi di Comunicazione, Arezzo.[8] The list of the loaned instruments was formulated as the assistant director visited the two collections on several occasions, taking hundreds of photographs. Giuseppe Tornatore himself chose the items for the workshop set, underscoring its crucial role in the movie as a place where the fraud is metaphorically assembled.

The very last sequence of the movie shows a restaurant in Prague with mechanical, indented-wheel clocks and other clock mechanisms, symbolically representing the gears wedging the auctioneer and the passing of time in a

---

8 These included nineteenth-century geodetic instruments, eighteenth- and nineteenth-century microscopes, nineteenth- and twentieth-century calculating machines, typewriters, acoustic instruments, phonographs and gramophones, musical instruments, galvanometers, Ruhmkorff induction coils and interrupters, complete telegraphic sets, late nineteenth- and early twentieth-century telephone apparatus, several 1920–1940s short-wave radios, an oscilloscope, boxes full of electric and electronic components (capacitors, resistors, etc.), an early twentieth-century and a rare 1960 movie camera, a wooden 1915–1920 juke-box, late nineteenth to early twentieth-century photographic cameras, a large electrostatic machine, an X-ray device, several magic lanterns, eighteenth- and nineteenth-century clocks, and more.

fruitless yet hopeful wait. Further research would be required to determine whether these machines are original or purpose-made for the film.

A Few Other Cases

Historic scientific instruments have 'acted' on the set of numerous movies, and the instances presented here represent cases for which the authors were able to find a little information. In *Gangs of New York* (2002) by Martin Scorsese, a 1863 police telegraph room was reconstructed with the aid of five complete two-way telegraphic stations loaned from the Museo dei Mezzi di Comunicazione. Initially intended as part of the background for the set arranged at Cinecittà (Rome), the stations took a more active role in the movie when Scorsese decided to put them to work. They were electrically connected on the sly, simulating a nineteenth-century electrical wiring that had originally functioned with Leclanché-type cells.

Further cases of loans of historical instruments can be found in television documentaries such as *Galileo's Battle for the Heavens* (NOVA Productions, 2002), where some sixteenth- and seventeenth-century astronomical and optical instruments were used to provide a credible appearance to Galileo's study, or the TV historical fiction *La Baronessa di Carini* (RAI, 2007), portraying original, early nineteenth-century geodetic instruments (theodolites, levels, drawing instruments) in scenes reproducing the campaign for drafting the topographic chart of the Kingdom of the Two Sicilies.[9] In both cases, historical instruments have contributed to reconstruct, as closely as possible, the ambience of that time, thus helping to render convincing the film's historical aspects, which can be, as already noted, one of their main roles on the set.

In the famous TV series *Longitude* (Channel 4, 2000), directed by Charles Sturridge, replicas of John Harrison's chronometers were used; information has not yet been obtained about other historical instruments used on the set. Original instruments were used for the first two Harry Potter films,[10] both by Chris Columbus: *Harry Potter and the Philosopher's Stone* (Warner Bros., 2001) and *Harry Potter and the Chamber of Secrets* (Warner Bros., 2002). In at least three other famous movies—*Tora! Tora! Tora!* (20th Century Fox, 1970) by Richard Fleischer, *The Name of the Rose* (Columbia Pictures, 1986) by Jean-Jacques Annaud and *Otello* (PIC, 1986) by Franco Zeffirelli—scientific instruments

9 In both cases the instruments came from Fausto Casi collection, Arezzo.

10 Silke Ackermann, personal communication.

appear on the set. These are just a few examples. To obtain a full survey of the use of instruments on movie sets still requires more in-depth research.

Conclusions

In 2010 the INAF—Palermo Observatory created an exhibition at the Specola Museum, trying to reconstruct the set used by Visconti for the sequences of *The Leopard* related to the Prince's private observatory.[11] The public could 'step inside' the reconstructed set and make comparisons with the movie sequence, which was being projected in a nearby space. This successful exhibition was highly appreciated by visitors as it inspired a different approach to the scientific instruments, creating an emotional relationship between people and the novel, as well as the movie. Although surveillance was strict, visitors were allowed to step into the set reconstruction and get close to the instruments.

The degree of success gained by this 'experimental' exhibition, the first of this kind, has confirmed that the public appreciate unusual displays and that interdisciplinary aspects from the humanities and the arts are very efficient means to facilitate the visitors' fruition of scientific instruments.

On the other hand, the exhibition of instruments used in movie sets, in the absence of an appropriate context, seems to be a less efficient way to highlight them. The replica of the Galilean telescope used by Giuseppe Tornatore, although displayed in the shop window of the Associazione Culturale "Il Mondo in Casa" in Arezzo, together with the relevant information on its loan for the movie, did not produce an increased number of visitors and neither did it enhance the general degree of interest for the item. This was probably due to the lack of an appropriate interdisciplinary context that could have amplified visitors' perception of the instrument and added some evocative meaning.

The above considerations may provide some initial suggestions and themes for future larger exhibitions on the role played by scientific instruments in movie sets. The purpose of this article indeed was to highlight two major cases and thereby to draw attention to this fascinating and unusual subject. Since it has been difficult to obtain information on the wider topic, perhaps this discussion might open up new sources of evidence concerning instruments on movie sets.

---

11 See Chinnici, Randazzo, op. cit. (n. 2), pp. 11–12.

CHAPTER 12

Display of Instruments on Seventeenth-Century Astronomical Frontispieces

Inga Elmqvist Söderlund

Introduction

This essay explores how the illustrations in seventeenth-century astronomical books address consumption. It asks how instruments are displayed on frontispieces and illustrated title pages. Illustrations on the first pages of the book, when there are any, advertise the content. This was where the producers of the book could communicate to the user why it should be bought, read, and further used. I am particularly interested in the kinds of settings in which the instruments are displayed, in the human actions to which they are connected, and in how these images are referred to in the text, particularly as concerns consumption.

Frontispieces and Illustrated Title Pages

I have studied the motifs of a great number of astronomical frontispieces and illustrated title pages that were individually designed for the title in the seventeenth century. I made a survey of 291 different motifs comprising astronomy in a wider sense. After astronomical phenomena, scientific instruments were depicted most commonly. Of the books in my sample, 62 percent show one or several scientific instruments. If we take into consideration only those books that have a frontispiece or illustrated title page where the illustration has a more prominent role (illustrations occupying a whole page), the proportion of illustrations that include instruments is as high as 86 percent.[1] Besides instruments, the other main consumable product on display is books. The scientific instrument as such seems to be a specific attribute of astronomy. Astronomical instruments could also be used to signify the mathematical sciences in general. Contemporary theoretical works could state this. For example, the Jesuit Claude-François Menestrier in a book on the theory of emblems advised that

1 Inga Elmqvist Söderlund, *Taking Possession of Astronomy*, Stockholm, 2010, p. 122.

 | DOI 10.1163/9789004264403_013

the appropriate attributes for depicting mathematics were globes, spheres, astrolabes, proportional circles and telescopes.[2] That was a more general advice for those who wanted to invent emblems; but in the case of most astronomical books, the instruments depicted are specialized for the content of the text.

Scientific instruments and books were often sold together in shops or by the same agents. Many authors of astronomical books were also suppliers of instruments. The books provided the possibility to market existing instruments or those that were only imagined but not yet constructed in brass and glass. Instruments and books were often collected, displayed and used in the same sites, such as in the library or the *Kunstkammer*. This strong relationship between books and scientific instruments is also visualized in the many printers' emblems that depict scientific instruments (armillary spheres, compasses, globes, etc.). One such famous emblem is the one used in variations by the Blaeu family of printers and mapmakers (*Fig. 12.1*). It depicts centrally an armillary sphere; to its left is a personification of Saturn or Time, with his scythe and hour glass, and to the right we see Hercules slaying the Hydra with a club, obvious references to antiquity. Below the text the phrase "*Indefessus Agendo*" (indefatigable action) alludes to industriousness. This well-know printer's mark communicated to the user that this was a book made by an established printing-house, and advertised a certain quality which could be recognized. As this emblem was used in a number of titles, it did not comment on anything specific written in the book. This use of specific images is not unlike modern branding of consumer wares.

Some frontispieces that are individually designed for the title shows similar motifs. A central armillary sphere with a man at each side is found on the frontispiece of Georg Welsch's, *Commentarius in Ruznameh Nauruz* (*Pl. 13*).[3] In this case the design alludes to the content of the book. The exotic-looking men are wearing turbans as the book is about Turkish and Persian chronology. The inscriptions on the pedestal are accordingly in Persian and Arabic. The conspicuous central location of the armillary sphere and its elevation on a pedestal stresses preciousness and singularity of the artefact. The attention of the beholder is directed towards the armillary sphere by a number of pointers in the image. From above the hand of God indicate where the beholder should look. Also lines of the design guide the eyes of the beholder. The rod of the man to the left makes a line in the direction of the armillary sphere, and similarly the arm of the man to the right makes a line in the direction of the armillary

2 Claude François Menestrier, *L'Art des emblemes*, Lyon, 1662, p. 68.

3 Georg Welsch, *Commentarius in Ruznameh Nauruz sive Tabulae aequinoctiales novi Persarum & Turcarum anni*, Augsburg, 1676.

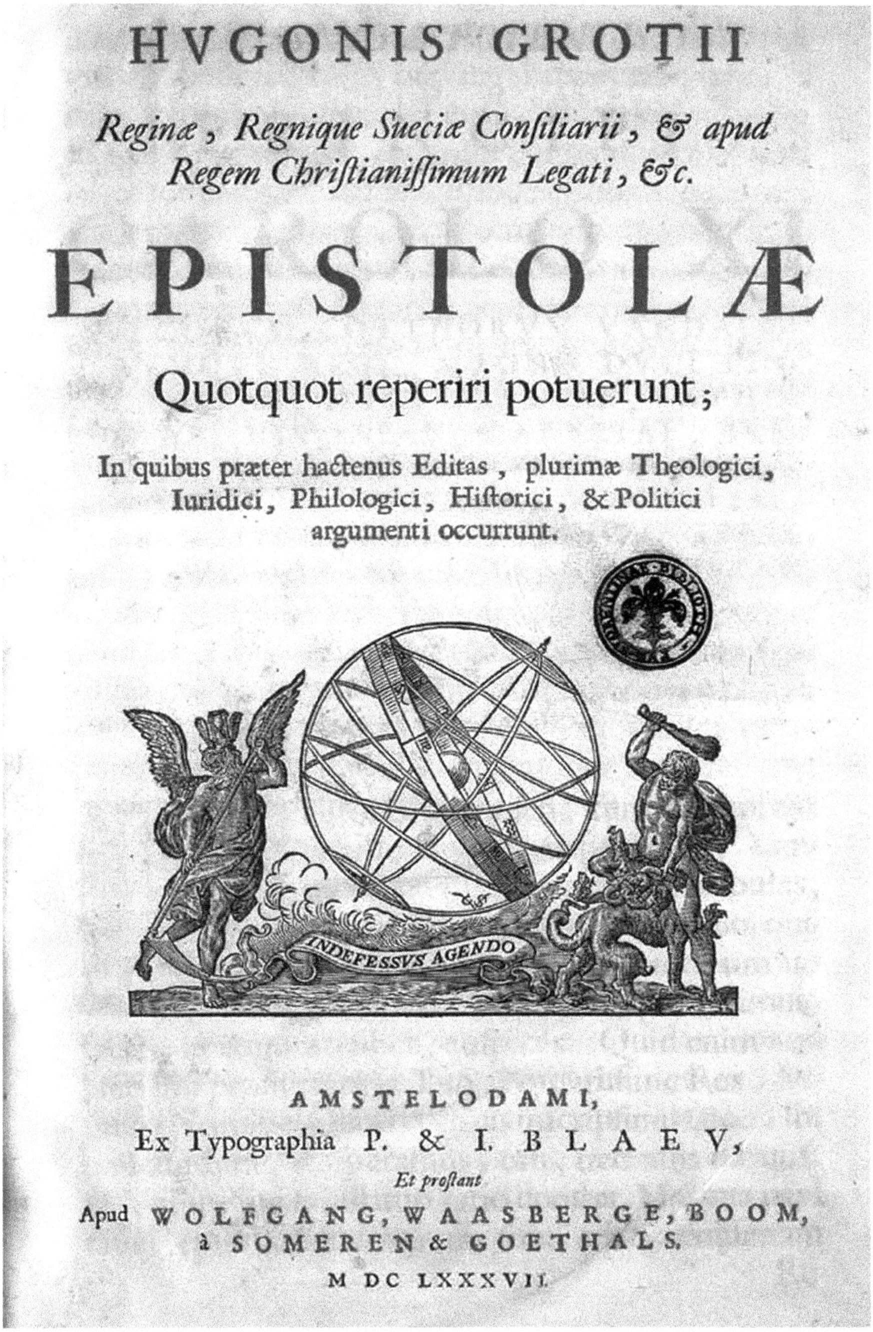

HVGONIS GROTII

Reginæ, Regnique Sueciæ Consiliarii, & apud Regem Christianissimum Legati, &c.

EPISTOLÆ

Quotquot reperiri potuerunt;

In quibus præter hactenus Editas, plurimæ Theologici, Iuridici, Philologici, Historici, & Politici argumenti occurrunt.

INDEFESSVS AGENDO

AMSTELODAMI,

Ex Typographia P. & I. BLAEV,

Et prostant

Apud WOLFGANG, WAASBERGE, BOOM, à SOMEREN & GOETHALS.

M DC LXXXVII.

FIGURE 12.1 *Hugo Grotius,* Epistolae, *Amsterdam, P. and I. Blaeu, 1687: title page with the emblem of the Blaeu printing establishment*
(BNCF, FLORENCE, AND MUSEO GALILEO DIGITAL LIBRARY, FLORENCE).

sphere, as well as his gaze, which further leads the eyes of the beholder towards the armillary sphere. The man to the left holds a globe, and his pointer is not only in the direction of the armillary sphere, but he demonstrates the passing of time by pointing at his globe on one of the signs of the Zodiac (Aries). The other man is taking the altitude with a quadrant. The three types of instruments—the armillary sphere, the globe and the quadrant—were not only practical for chronology but also coveted collectables, and it is here alluded to that they carry an exotic flavour, something especially appreciated by collectors. There are numerous examples of frontispieces that show instruments being admired or venerated, placed high on a pedestal, garlanded with flowers, or encircled by putti. The sites in which the instruments are placed are often a room with a great many artefacts indicating a larger collection, possibly in a library or study, and sometimes outside, a balcony or a garden, a suitable site to stage instruments in order to display the refined taste and education of the owner.

Conspicuous Consumption

The study of commercial aspects and consumption of early modern collecting and science seems to be a fruitful perspective, and a number of researchers have pursued these questions within the studies of material culture. Lisa Jardine laid out the consumption of both physical goods and knowledge in Renaissance Europe.[4] For a somewhat later period, which is relevant for this article, Linda Levy Peck investigated consumption and demand of luxury goods in seventeenth-century England. She highlighted a number of fields of consumption, such as personal accessories, textiles and paintings. She pointed out that both scientific knowledge and scientific artefacts were in demand as part of conspicuous consumption.[5] Scientific artefacts were generally a field for male collecting, but as a recent study shows, such collecting could be pursued as a princely virtue also by a woman.[6] Adam Mosely further showed in his study of Tycho Brahe and his network that commercial aspects were relevant to astronomers in late sixteenth century Europe.[7] It appears that knowledge,

4 Lisa Jardine, *Worldly Goods: A New History of the Renaissance*, London, 1996.

5 Linda Levy Peck, *Consuming Splendour*, Cambridge, 2005, pp. 311–345.

6 Lisa Skogh, *Material Worlds: Queen Hedwig Eleonora as Collector and Patron of the Arts*, Stockholm, 2013, pp. 225 ff.

7 Adam Mosley, *Bearing the Heavens: Tycho Brahe and the Astronomical Community of the Late Sixteenth Century*, Cambridge, 2007, pp. 209–288.

scientific artefacts, and books, at least at the higher end, were intended for a European rather than local audience. This is particularly so in the case of books published in Latin. Giles Mandelbrote identified a few groups of consumers of books, scientific instruments and natural curiosities in Europe. He pointed out that it was not only the learned who bought these things, but also interested amateurs and scientific virtuosi, who used them to help define who they were.[8]

If we accept that commercial interests were relevant in the production of knowledge as well as of scientific artefacts, it is not unreasonable to expect to find evidence of this in different media. One available means was illustrations in books, where products could be staged in ways to promote consumption.

On the illustrated title page of Philips van Lansbergen's *In quadrantem* the instruments are shown as fashionable accessories (*Fig. 12.2*).[9] Two young men are depicted using quadrants knowledgably. The one in the front is taking the altitude of the sun, and the man behind is measuring the church-tower Lange Jan in Middleburg in the background. The man in the front seems to be showing off his fashionable clothes, with his plumed hat, ribboned shoe and knee positioned so that it can be viewed from its best vantage point. Further instruments are depicted: a quadrant is laid out on the ground and an astrolabe is depicted in the top left corner. In common with modern advertisements where products are featured, these two instruments are shown in such a way that their identifying features are clearly visible. In order to display these features the perspective of the quadrant on the ground has been distorted, and it appears to be depicted at a very odd angle, whereby the instrument can be seen from the side. Here the objects are transformed from being impersonal commodities and are staged to feature a particular moment in an ideal man's life. The image shows how practical astronomy could be performed elegantly with the right possessions. Astronomy is depicted as a delightful pastime, or as conspicuous leisure, in the tradition of the liberal arts, when these were pursuits of leisure for those who were liberated from having to work hard physically in order to survive. The quadrant and astrolabe are here presented as means with which the user could fashion himself with attributes and appropriate actions to display his learning and taste.

---

8 Giles Mandelbrote, "Scientific Books and their Owners: A Survey to ca. 1720", in Andrew Hunter (ed.), *Thornton and Tully's Scientific Books, Libraries and Collectors*, Aldershot, 2000[4], pp. 333–366: 336.

9 Philips van Lansbergen, *In quadrantem tum astronomicum tum geometricum nec non in astrolabium introductio*, Middelburg, 1635.

FIGURE 12.2 *Philips van Lansbergen,* In quadrantem…, *Middelburg, 1635: illustrated title page, attributed to Adriaen van de Venne*
(LIBRARY OF THE ROYAL SWEDISH ACADEMY OF SCIENCES AT STOCKHOLM UNIVERSITY LIBRARY. PHOTOGRAPH: STOCKHOLM UNIVERSITY LIBRARY).

The concept "conspicuous consumption" was coined by the economist Thorstein Veblen in his 1899 book *Theory of the leisure class*. He pursued the relationship between consumption and the definition of social identity and hierarchy. The book is a product of its time, and the author uses theories of social evolution which now feels outdated, but Veblen was onto something which is still current as a grid of interpretation: consumption and its relationship to social identity. Veblen suggested that actions carry significant social meaning, in that "conspicuous consumption" gratifies the consumer with a symbolic gain.[10] The concepts of "conspicuous consumption" and "conspicuous leisure" have since been inspirational for a number of important studies.

We might question to which extent it is possible to shape your own identity and to which extent it is given. In an anthropological sense identity is partly given, that is essential, and partly defined by outer circumstances. In repeating certain behaviours, to repeatedly show yourself from the same side it is possible to enforce your identity and to become more like your ideal (i.e., desired) self. This is the differentiated identity, the identity which is less stable, which can be acquired. To own a certain style can be experienced as positive, to have good knowledge of yourself, to be consequent. Shaping your own identity can comprise everything from how you express yourself, the clothes, the hairstyle, the acquisition of objects, but also the acquisition of knowledge. Pierre Bourdieu believed that the tastes that influence such consumption are socially conditioned and can be used as strategies to display belonging to or separation from various social groups and class.[11] Stephen Greenblatt used the term "self fashioning" in his study of sixteenth-century England, through the interpretation of literary works, where he sees human identity as the result of both social and individual shaping.[12] The particular role of artefacts was further explored in his book *Marvelous Possessions*.[13]

It is not reasonable to address consumption without mentioning expenditure. The acquisition of artefacts was linked to financial wealth and power. Artefacts could be produced by the transformation of materials, bought, exchanged, presented as gifts, become inherited, or taken as war-booty; and in whatever form acquired, they represented a monetary value. Many scientific instruments were expensive, particularly if beautifully crafted or if made from precious materials. Cheaper versions could be produced by printing templates on paper, where the buyer himself could colour, cut out and paste the sheet onto a piece of wood. Prices could vary over time. Telescopes were very expensive at the beginning of the century but then gradually became cheaper. For example, in Augsburg around 1650 a large telescope cost about as much as two good horses.[14]

---

10 Thorstein Veblen, *Theory of the Leisure Class: An Economic Study in the Evolution of Institutions*, New York, 1899, pp. 68 ff.

11 Pierre Bourdieu, *Distinction: A Social Critique of the Judgement of Taste*, Cambridge (Mass.), 1984, pp. xxiv–xxx.

12 Stephen Greenblatt, *Renaissance Self-Fashioning from More to Shakespeare*, Chicago, 1980.

13 Greenblatt, *Marvelous Posessesions*, Oxford, 1991.

14 Inge Keil, *Von Ocularien, Perspicillen und Mikroskopen, von Hungersnoten und Friedensfreuden, Optikern, Kaufleuten und Fürsten: Materialien zur Geschichte der*

A few frontispieces in fact depict the coins. A treasury-chest filled with coins can be seen on the frontispiece of Mario Bettini's *Aerarium Philosophiae Mathematicae* (*Fig. 12.3*).[15] In the illustration, to the left, a young well-clad man has removed his hat, slightly bowing his head and is gesturing towards a chest filled with coins. In front of the chest stands erect an elderly Jesuit, at once pointing with one hand to the mathematical garden and palace spreading its wonders behind and with his other gesturing towards the chest of coins as if collecting a fee. As part of the architectural backdrop on the left there are mathematical sculptures in female form, carrying a number of instruments and models. The sculpture in the front left is carrying lenses through which rays of light are refracted. In the loggia and the garden a number of young men seem to be enjoying the mathematical pleasures at leisure, while making and discussing instruments accompanied by music. The garden presents beauty and intellectual challenges, where men can behold, make, use, and discuss instruments in a leisurely and pleasant fashion. It presents an ideal place for intellectual exchange and enjoyment of conspicuous leisure.

On the famous frontispiece of Kepler's *Tabulae Rudolphinae* the Habsburg eagle is spreading coins over the temple of astronomy (*Fig. 12.4*).[16] The eagle is seen hovering over the temple with a coin in its beak, and coins are raining down over the famous legendary men disputing astronomy: Ptolemy, Hipparchus, Tycho, Copernicus, an unnamed Babylonian astronomer, and Aratus invisible as he is hiding behind a column, and also to Kepler sitting at a desk on the base to the left, and the printers to the right. The central argument is between Tycho and Copernicus on the geo-heliocentric and heliocentric models. This frontispiece is accompanied by a long hexameter poem, the "*Idyllion*", inserted after the title page, written by the Latinist Johannes Hebenstreit. This poem allows us to decipher some of the enigmatic features of the image. I will not refer to all its details as this has already been done elsewhere.[17] Here it shall suffice to highlight the instruments as they presented in conspicuous display.

---

optischen Werkstatt von Johann Wiesel (1583–1662) und seiner Nachfolger in Augsburg, Augsburg, 2003, p. 94.

15 Mario Bettini, *Aerarium philosophiae mathematicae*, Bologna, 1648.

16 Johannes Kepler, *Tabulae Rudolphinae*, Ulm, 1627.

17 Stefano Gattei, "The Engraved Frontispiece of Kepler's *Tabulae Rudolphinae* (1627): A Preliminary Study", *Nuncius* 24, 2 (2009), pp. 341–365; Arwed Arnulf, "Das Titelbild der *Tabulae Rudolphinae* des Johannes Kepler. Zu Entwurf, Ausfuhrung, dichterischer Erlauterung und Vorbildern einer Wissenschaftsallegorie", *Zeitschrift des Deutschen Vereins für Kunstwissenschaft* 54/55 (2000–2001), pp. 176–198; Franz Hammer, "Nachbericht", in Kepler, *Gesammelte Werke*, Munich, 1938–2009, 21 vols.: v. 10, pp. 22*–27*; Elmqvist Söderlund, op. cit. (n. 1), pp. 209–211.

FIGURE 12.3 *Mario Bettini,* Aerarium philosophiae Mathematicae, *Bologna, 1648: Frontispiece by Francesco Curti* (NATIONAL LIBRARY OF SWEDEN).

FIGURE 12.4 *Johannes Kepler,* Tabulae Rudolphinae, *Ulm, 1627: frontispiece by Georg Celer* (LIBRARY OF THE ROYAL SWEDISH ACADEMY OF SCIENCES AT STOCKHOLM UNIVERSITY LIBRARY. PHOTOGRAPH: STOCKHOLM UNIVERSITY LIBRARY).

On the top of the temple a number of goddesses, Urania and her servants, are carrying instruments as attributes. To the left, instead of the Sun, the head of the goddess *Physica Lucis* emanates rays all around and grasps at a sphere which can demonstrate principles of the spherical Moon. Next is *Optica* holding a telescope, presented as superior to anything the culture of antiquity could produce. To her right is *Logarithmica* with rods, one twice the length of the other. To her right is *Doctrina Triangulorum* with a set square and a pair of compasses; further to the right *Stathmica* holds a balance. The goddess furthest to the right is *Magnetica*, with loadstone and a magnetic needle. Below in the temple of which the architecture indicates the progression of astronomy, a number of instruments are displayed. We see an armillary sphere, a celestial globe, cross staff, a parallactic ruler, a quadrant, a sextant, an astrolabe and a dial for the Metonic cycle. The instruments tend to be displayed next to the astronomer who used that kind of instrument, except for the Babylonian astronomer who did not have access to any instrument except his fingers with which he takes the altitude. In the image is a mixture of modern, ancient and exotic. The illustration shows what instruments would be proper to collect, and what subjects would be befitting to be able to converse on. The "*Idyllion*" mentions the instruments several times, especially those of Tycho. It is imperative in directing attention to how the depicted instruments are displayed. It says: "Do you see how Tycho's instruments, casted from golden metal, shine as eternal monuments?".[18] These instruments were famous throughout Europe, not least due to the descriptions and illustrations in the *Astronomiae Instauratae Mechanica*.[19] They were admired for their scientific value, their beauty and, not least, their costliness.

Johannes Hevelius used Tycho's book on instruments as a role model for his own *Machina Coelestis*. The latter is filled with descriptions and attractive illustrations of various types of instruments. In the introduction Hevelius writes that the student of astronomy must not be deterred from "fear of the expenses".[20] Further he states that not everyone could build and equip an observatory. The cost necessarily required a prince, a great patron, who could

18 Kepler, *Gesammelte Werke*, cit. (n. 17), v. 10, p. 17. See also Elmqvist Söderlund, op. cit. (n. 1), p. 371.

19 Tycho Brahe, *Astronomiae Instauratae Mechanica*, Wandsbek, 1598.

20 Johannes Hevelius, *Machina Coelestis pars prior*, Danzig, 1673, p. 22 (translated by Peter Sjökvist). See Elmqvist Söderlund, op. cit. (n. 1), p. 376.

support the activity.[21] In this respect Hevelius himself was very successful in attracting a great deal of support from patrons. These exclusive books dedicated to princes situate astronomy as a princely pursuit which is shown both in the texts and the illustrations. Needless to say, the books were a good opportunity to market the knowledge, services and products that the astronomer could supply to prospective clients.

A great number of frontispieces depict imaginary spaces, as in Kepler's temple of astronomy. The illustrations show worlds that are idealised, spaces for dream, myth, imagination, grandeur, and the exotic. These concepts have been linked to modern mass consumption. In advertisements and the display of merchandise dream worlds can be created. The mythical land of plenty is created within the consumer's mind.[22] The producers of the advertisement try to create a medium with which to make a link, a mental association, between the consumable goods and romance, something remote from everyday experience, something imaginative or suggestive of grandeur and passion, of being exotic, imaginative and idealized. Colin Campbell specified values that are often alluded to in modern advertising as "'remote from everyday experience', 'imaginative' or suggestive of 'grandeur' or 'passion' [...] the pictures and stories used are often typically 'romantic' in the broader sense of being exotic, imaginative, idealized; whilst the very purpose of advertisements, of course, is to induce us to buy the products which are featured: in other words to consume".[23] These concepts, found in modern advertising of consumer ware have, as I have argued, also appeared on frontispieces in relation to scientific instruments. There is however one major difference, which is valid for a significant portion of the frontispieces, *viz.*, the amount of time and effort expected from the beholder in order to make sense of the pictorial programme. Many frontispieces require an engaged, intensive, and extensive consideration which is different from the superficial beholding presupposed for modern product placement.

Enigmatic emblems, and images filled with associations to ancient learning were generally widely fashionable in the seventeenth century. In the astronomical context they identify the books as thoroughly erudite. Compared to other illustrations in the book, the frontispieces tend to contain more abstract

---

21 Hevelius, op. cit. (n. 20), p. 419.

22 Rosalind H. Williams, *Mass Consumption in Late Nineteenth-Century France*, Berkeley and Los Angeles (California), 1982, pp. 211–384.

23 Colin Campbell, *The Romantic Ethic and the Spirit of Modern Consumerism*, Oxford, 1987, pp. 1–2.

concepts. Most illustrations are not signed, and often the artists remain unknown. Often a team would be involved, and usually the more famous artist designed the frontispiece rather than the illustrations accompanying the text. Hevelius's *Machina Coelestis* can serve as an example. For the frontispiece the artist Adolf Boy and the engraver Jeremias Falck were commissioned (*Fig. 12.5*). The other artists who signed the numerous illustrations were all accomplished but less famous and skilled than Boy and Falck. Boy could make the invention more learned and Falck was the superior craftsman in engraving. The frontispiece is explained in the foreword, and many of the details have profound meanings related to the content of the book, and to the ideal qualities of the astronomer, which presupposes an intensive and thoughtful beholder.[24] For example, each detail on the obelisk is explained. It starts out: "The first emblem, going upwards, shows the skull of a man, with the inscription: *Acutissimo* [with very sharp head], as if said: a true astronomer should have a very bright intellect and a very sublime mind, in order to discern abstruse circumstances without difficulty, to explore them completely and to penetrate them".[25] Here the physical and mental qualities of the astronomer are described, and also the attributes of an armillary sphere, a sextant, a celestial globe, and books are displayed. The astronomers (Hipparchus, Ptolemy, Copernicus and Tycho) are shown as actors on the stage. In fact a Roman theatre is depicted in the background. The impression of the theatre is strengthened by a stage coach descending from above with the personifications at the astronomer's disposal. Here the message is quite clear that instruments and books are necessary for a proper astronomer and that knowledge is necessary in order to participate in a learned conversation.

This use of emblems and hieroglyphs with profound meanings is often specific for the frontispiece. Other illustrations may have more of a decorative or explanatory function. One of the vignettes which adorn the dedication to Louis XIV is more decorative. It shows an assemblage of instruments and putti, a beautiful depiction of the land of plenty of instruments (*Fig. 12.6*). The placing of the artefacts gives the impression of abundance and riches fit for a king, inviting the beholder to admire the beauty of the objects. The depicted instruments are related to the book: quadrants, sextants, telescopes, and a globe, but this illustration is not as extensively explained as the emblematic frontispiece.

24 See Elmqvist Söderlund, op. cit. (n. 1), pp. 156, 177 ff., 293 ff., 301 ff., and 376 ff.

25 Hevelius, op. cit. (n. 20), p. 21 (translated by Peter Sjökvist). See Elmqvist Söderlund, op. cit. (n. 1), p. 376.

FIGURE 12.5 *Johannes Hevelius,* Machina Coelestis pars prior, *Danzig, 1673: Frontispiece by Adolf Boy and Jeremias Falck*
(LIBRARY OF THE ROYAL SWEDISH ACADEMY OF SCIENCES AT STOCKHOLM UNIVERSITY LIBRARY).

FIGURE 12.6 *Johannes Hevelius,* Machina Coelestis pars prior, *Danzig, 1673: Illustration by Andreas Stech and Johann Benßheimer*
(LIBRARY OF THE ROYAL SWEDISH ACADEMY OF SCIENCES AT STOCKHOLM UNIVERSITY LIBRARY).

Further inside the book a great many illustrations also show instruments and their use. Yet these are not of the emblematic quality of the frontispiece. Hevelius's arsenal of instruments as well as his home and observatory are laid out in detail. One of the illustrations shows how Hevelius has set up a great telescope outside the city of Gdansk (*Fig. 12.7*). This illustration appears documentary of a real event, as described in the book, even if the scene is likely bettered by the artist. It shows how the display of a great telescope could draw large crowds and attention from prominent individuals. Telescopes could be useful artefacts both in promoting knowledge and in representation and in social interaction. Keeping and displaying collections could be a useful political tool.[26] In this case the sheer size of the instrument was likely to stir a conversation. It was perhaps befitting that in order to capture the grandeur of the heavens such a huge machine was needed. Hevelius's book is a real treasure trove, illustrating instruments in unusual richness. The illustrations in the book show what the instruments were for, but each category in its own way.

26 Thomas DaCosta Kaufmann, "Remarks on the Collections of Rudolf II: The *Kunstkammer* as a Form of *Representatio*", *Art Journal* 38, 1 (1978), pp. 22–28.

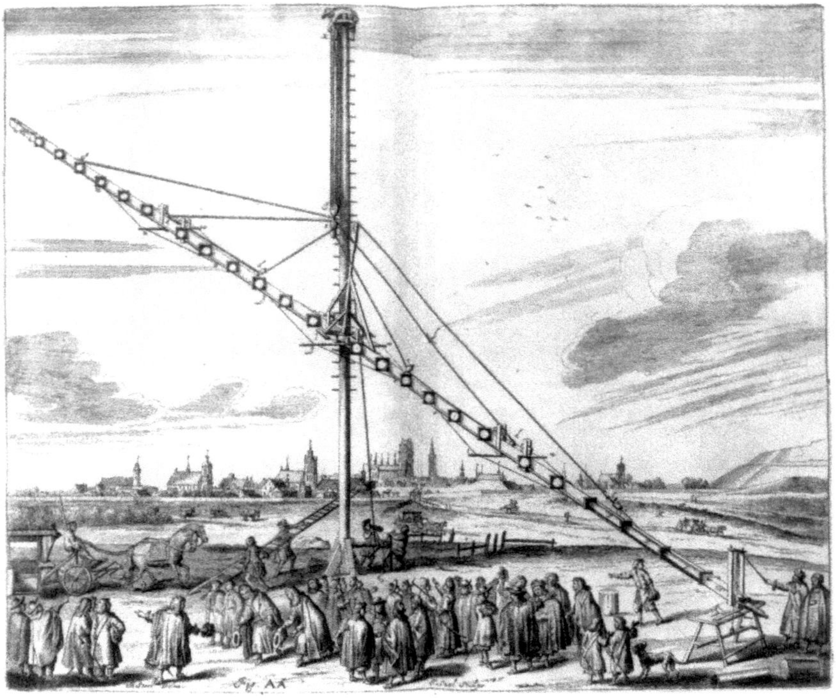

FIGURE 12.7 *Johannes Hevelius,* Machina coelestis pars prior, *Danzig, 1673: Illustration by Andreas Stech and Isaac Saal*
(LIBRARY OF THE ROYAL SWEDISH ACADEMY OF SCIENCES AT STOCKHOLM UNIVERSITY LIBRARY. PHOTOGRAPH: STOCKHOLM UNIVERSITY LIBRARY).

Conclusion

Frontispieces and illustrated title pages contributed to shape the identity of astronomy by suggesting that astronomical instruments were an appropriate attribute for a person who wanted to be associated with the field. Books and astronomical instruments were consumer goods that were most prominently displayed on astronomical frontispieces in the seventeenth century. The settings in which these products were staged are often idealized spaces such as an imaginary palace or a garden, but sometimes identifiable spaces are used. The imaginary spaces depict an enhanced reality, a pleasant dream world. In the illustrations, the consumable products are displayed with their distinguishing features in attractive arrangements or with beautiful persons in ideal poses in ideal places, sometimes with exotic details. In *Theory on the Leisure Class,*

Veblen wrote how a man of wealth should not consume indiscriminately: "In order to avoid stultification he must also cultivate his tastes, for it now becomes incumbent on him to discriminate with some nicety between the noble and the ignoble".[27] This understanding of the relationship between taste, consumption and social identity can be useful in the interpretation of the early modern display of scientific instruments. Instruments had the potential to provide the user with a symbolic gratification. The pursuit of the liberal arts, to which astronomy belonged, was a noble occupation in which only those who were liberated from heavy manual labour could indulge. The books could provide the man of wealth with the means to acquire cultivation. The illustration would show how to collect, use, display and talk about instruments, as well as create suitable settings for them. For the man of lesser means the illustrations could provide means for dreams. Collecting books on astronomy could substitute for collecting the instruments themselves. And books could be acquired at a lower price in which the limits were not financial but simply the imagination of the beholder.

---

27 Veblen, op. cit. (n. 10), p. 74.

Index